Smart Materials for Smart Devices and Structures

Selected, peer reviewed papers from the
Symposium C, European Materials Research Society, Fall Meeting,
Warsaw University of Technology,
September 15-19, 2008

Edited by

Marcin Leonowicz and Dariusz Oleszak

TRANS TECH PUBLICATIONS LTD
Switzerland • UK • USA

Trans Tech Publications Ltd
Laubisrutistr. 24
CH-8712 Stafa-Zurich
Switzerland
http://www.ttp.net

Volume 154 of
Solid State Phenomena
ISSN 1012-0394
(Pt. B of Diffusion and Defect Data - Solid State Data (ISSN 0377-6883))

Full text available online at http://www.scientific.net

***Distributed** worldwide by*

Trans Tech Publications Ltd
Laubisrutistr. 24
CH-8712 Stafa-Zurich
Switzerland

Fax: +41 (44) 922 10 33
e-mail: sales@ttp.net

and in the Americas by

Trans Tech Publications Inc.
PO Box 699, May Street
Enfield, NH 03748
USA

Phone: +1 (603) 632-7377
Fax: +1 (603) 632-5611
e-mail: sales-usa@ttp.net

Printed in the Netherlands

Preface

European Materials Research Society organises every year the Fall Meeting in September in Warsaw. The 2008 Meeting was held from 15 to 19 September at the Warsaw University of Technology. This volume of Solid State Phenomena contains proceedings of the Symposium C *Smart Materials for Smart Devices and Structures*

Smart Materials are recognised as materials that can respond to environmental stimuli with particular changes in some variables such as mechanical and physical properties, structure, composition or functions. Growing interest in their development is driven by emerging applications and integrations of smart materials into industrial systems including civilian, industrial, medical and military applications. Among them are composite multiferroic materials exhibiting two or more ferroic features such as ferromagnetism/magnetostriction, ferroelectricity/piezoelectricity or ferroelasticity/shape memory effects due to their unusual responses, including very large magneto-electric susceptibility, giant magnetostriction and energy coupling coefficients approaching one. The other systems include shape memory and magnetic shape memory alloys, magnetostrictive materials, magnetorheological fluids and polymers.
The Symposium highlighted advances in the development of smart materials with a special focus on numerical modelling, fabrication methods, technologies, and properties, sensing and processing capabilities, in the light of their potential applications in smart structures, devices and systems.
Among the relevant scopes of the Symposium were new capabilities, remarkable performance enhancement or cost reduction, new ideas and directions for future development.
The Symposium brought together researchers from various fields, including contributions from materials engineering, mechanical engineering, theoretical and applied mechanics, electronics, electromechanics and manufacturing.

We want to take the opportunity of thanking everyone involved in the organization of this meeting: the Local Organizing Committee, the International Scientific Committee, the Reviewers and especially all the Participants whose contributions assured the high scientific level of this Symposium.

Chairman of the Symposium
Professor Marcin Leonowicz

Smart Materials for Smart Devices and Structures

Table of Contents

Solid State Phenomena Vol. 154 (2009) pp 1-7
online at http://www.scientific.net

Rapid prototyping methodologies for ceramic micro components

P. R. Chalker[a], K.M. Berggreen[b], A. T. Clare[c], J. Singh[d], and C.J. Sutcliffe[e]

University of Liverpool, Department of Engineering, Brownlow Hill, Liverpool, L69 3GH, U.K.

[a]pchalker@liv.ac.uk, [b] k.m.berggreen@liverpool.ac.uk, [c]a.t.clare@liverpool.ac.uk, [d]jetinder.singh@liverpool.ac.uk, [e]c.sutcliffe@liverpool.ac.uk

Keywords: Rapid prototyping, Selective laser sintering, $BaTiO_3$, Piezoelectric.

Abstract. This paper investigates the rapid prototyping of components from ceramics. Selective laser sintering of barium titanate powder has been investigated as a potential route to manufacturing freeform electro-ceramic based components. Raman spectroscopy has been used to investigate the effect of laser power and oxygen partial pressure on the phase and composition of the sintered material. It is shown that the sintering laser power strongly influences the composition. As the laser power is increased during sintering, the E_g mode of anatase becomes evident at 640 cm^{-1}. When the oxygen partial pressure in the sintering atmosphere is reduced from 19% to 0.2% the whole depth of the powder bed sampled by Raman spectroscopy resembles bulk anatase.

Introduction

The continual shortening of product life cycle has driven the demand on manufacturers to reduce the time to market of their products. In the late 1980s, a class of manufacturing processes known as 'rapid prototyping' emerged that sought to reduce the time-to-market. Drawing on expertise from a range of disciplines, these technologies have enabled product designers to produce prototypes directly from computer aided design (CAD) data in a matter of only a few hours. Since then, the rapid prototyping industry has seen continued growth and an increasingly diverse range of industries are using the technologies for applications in medicine, tooling and electronics. The majority of rapid prototyping or RP technologies use an additive fabrication method for producing components. One such technique, selective laser sintering (SLS) enables the final product to be made directly and rapidly, without tooling or additional machining. The scope of SLS is demonstrated by recent investigations as diverse as the fabricating of customized hydroxyapatite - filled polymeric composite implants [1] to the consolidation of electro-ceramic polymer-composite thin films on silicon substrates [2]. In each of these examples, sintering of the ceramic is facilitated by the presence of a polymeric matrix. The rapid prototyping of lead zirconium titanate (PZT) piezoelectric ceramics using this technique has been reported [3]. However SLS was used to fabricate sacrificial plastic molds of the desired structure, not for sintering the ceramic part itself. Subsequently an aqueous PZT suspension was cast in the mold and solidified to form a three-dimensional network gel.

The direct selective laser sintering of electro-ceramic materials is of significant interest. For instance, ferroelectric thin films as well as bulk ceramics have had a strong impact in a large number of technological applications such as in ferroelectric memories and piezoelectric devices. One of the most important ferroelectric materials, barium titanate ($BaTiO_3$), has been researched intensively. It possess a variety of useful properties that have potential to be exploited in thin film devices, such as non-volatile memory devices, electro-optic devices, thin film capacitors, and pyroelectric detectors [4-6]. Consequently, investigation of the preparation and characterization of $BaTiO_3$ thin films in recent years has been reported widely. A variety of methods, such as sputtering [7,8] metalorganic chemical vapor deposition [9,10] and pulsed laser deposition (PLD) [11-13] have been applied to prepare such films.

This study addresses the selective laser sintering of $BaTiO_3$ powder beds and the influence of process parameters on composition and microstructure. In the SLS process, a succession of individual part cross sections are created from powdered materials by sintering them with an ytterbium fibre laser. A layer of the build powder is pre-deposited and spread uniformly by a wiper blade. The laser 'writes' the layer pattern in the powder bed sintering the material and joining it to the previously formed build structure. Using this approach, SLS offers the potential to realise true high aspect ratio 3D structures, however SLS has not yet found its way into the mainstream of MEMS fabrication technologies, which is the motivating factor for the present work.

Experimental Details

SLS is a layer-based solid freeform fabrication process. The apparatus used is the MCP Realizer, which is a commercial SLS workstation with a 50 W continuous wave ytterbium fibre laser (IPG, Germany) operating with a wavelength of 1068-1095 nm. The scanning system used is a dual axis mirror positioning system (Cambridge Technology) and a galvanometer optical scanner, which direct the laser beam in the x and y axis through an F-theta lens. The variable focussing optics are Sill 300 mm focal length lenses, which produce a focussed beam spot size of 35 μm diameter at 50 W power. The sintering process was performed in an argon ambient, to which controlled concentrations of oxygen could be added via a mass flow controller. The build geometry is controlled using propriety Realizer software. The control software was programmed to write small discs with a diameter of 5 mm in the $BaTiO_3$ powder bed at different laser powers ranging from 1 to 20W, while the laser dwell time remained in the range 200 μs to 300 μs.

Characterisation of the laser processed $BaTiO_3$ thin films was performed using Raman spectroscopy, and scanning electron microscopy (SEM) with energy dispersive X-ray microanalysis (EDX). The micro-structures of the surface and the cross section of the laser processed films were investigated by SEM and EDX using a Hitachi Model S-4000 electron microscope. Raman spectra were excited using 514.5 nm radiation from an argon ion laser. The power impinging on the sample was kept at 17 mW. A Jobin-Yvon LabRam spectrophotometer with subtractive pre-monochromators coupled to the third spectrograph / monochromator with a 1800 grooves mm^{-1} grating was used for recording the spectra. A Peltier-cooled charge-coupled device (CCD at 80 K) detector was employed to detect the Raman signals. All of the spectra were recorded in backscattering geometry with a 50x microscope objective.

Results and Discussion

The feedstock powder has to flow easily when the powder is delivered to the build platform. Consequently the powder characteristics of shape and size are important in the SLS process. A trade-off between particle sizes large enough to ensure good flow-ability but small enough to ensure that the finished part has satisfactory tolerances has to be reached. For very small particles, Van der Waals attraction causes cohesive agglomeration which is negligible on a larger scale but inhibits the flow of finer powders. Particles with smooth surfaces flow more readily within the powder bed leading to a higher 'loose' density, consequently the solidified part is likely to have a higher density. Particle shape is also an important factor as symmetrical particles pack more efficiently and consequently spherical grains are preferred. These surface and granulometry parameters of the powder are closely related to the powder material and its production process. The powder used was characterised as spherical particle sizes in the range shown in fig. 1. The $BaTiO_3$ powder used in this work was supplied by Strem Chemicals, (99%; m.p. 1620°C; density 5.58 gcm^{-3})

A matrix of the critical process parameters, such as laser power, scan speed, sintering ambient, was devised in order to identify an SLS process for the sintering and crystallization of $BaTiO_3$ whilst preventing cracks or unwanted decomposition. The laser power and scan speed determines the

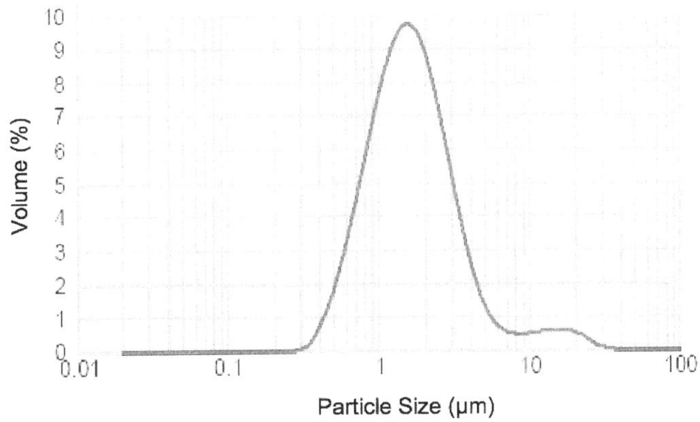

Fig. 1. $BaTiO_3$ particle size distribution used for selective laser sintering.

energy density input and consequently the temperature in the heated zone; the sintering depth; the crystallization kinetics; and the thermal shock.

A library of sintered samples was made on a single bed of $BaTiO_3$ powder. The laser power was systematically varied between 1 and 20 W; the laser scanning speed was varied between 0.1 and 0.2 ms^{-1}; and finally the level of oxygen in the argon ambient was adjusted between 0.2% (background) and 19%. The bed of sintered samples was analysed using Raman microscopy. The technique is sensitive to the phase and composition of the sintered material. At temperatures below 393 K $BaTiO_3$ is ferroelectric with the P4mm (C^1_{4v}) structure, which transforms to orthorhombic and rhombohedral structures at 278 and 183 K, respectively. Barium titanate is cubic above 393 – 408 K with the space group Pm3m(O^1_h).

Fig. 2 shows the Raman spectra of the sintered $BaTiO_3$ as a function of incident laser power. For these samples the oxygen content in the argon ambient was fixed at 19% coupled with an exposure time of 200μs and a scan speed of 0.125ms^{-1}. The peak intensities observed at 270, 308, 525 and 725 cm^{-1} are attributed to the A1(TO_2), E(TO_2), A_1(TO_3) and A_1(LO_3) of barium titatate modes of the room temperature P4mm phase. The asymmetric A_1(TO_3) mode couples weakly with the A_1(TO_2) mode [14,15] and the phonon wavevector directions of are randomly distributed with respect to the crystallographic axes in ceramics. This allows mode mixing and quasi-phonon modes in the spectrum which are sensitive to the sizes of crystallite domains within the sample. The Raman peak widths and positions remain unchanged at each laser power. This is indicative of the formation of large sintered grains, much larger than that of the feed-stock material.

As the laser power is increased during selective laser sintering a broad new feature centred about 639 cm^{-1} is observed to grow in intensity. This is assigned to the E_g mode of TiO_2 in the anatase polytype. The presence of the titania suggests the loss of barium oxide from the $BaTiO_3$. The exact mechanism of this decomposition process and where the barium oxide is removed to is not unequivocal. The melting temperature of barium titanate is 1620°C and it is plausible that the surface of the powder bed is reduced to a metallic film in an oxygen difficient environment.

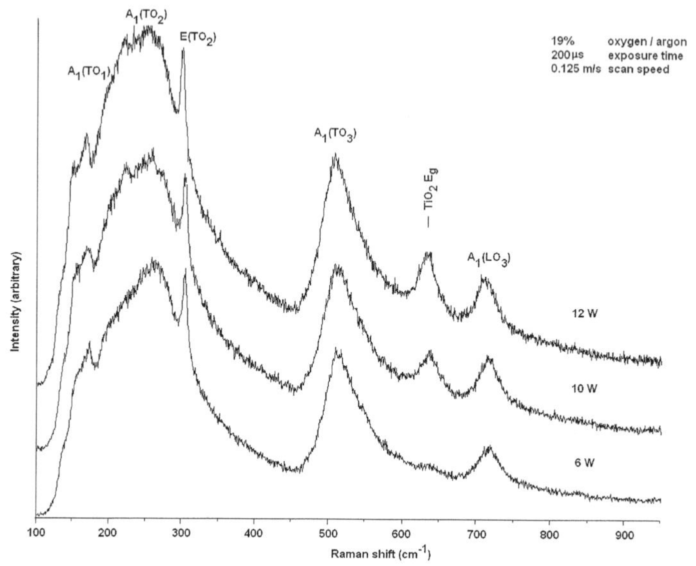

Fig. 2. Raman spectra of sintered $BaTiO_3$ as a function of incident laser power.

To investigate the role of oxygen partial pressure in stabilising the $BaTiO_3$ during sintering Raman spectra were recorded from samples prepared at O_2 fractions of 0.2 to 9% in argon. The incident laser power was held at 8 W coupled with an exposure time of 300μs and a scan speed of $0.1ms^{-1}$ Figure 3 shows the resulting Raman spectra of the sintered $BaTiO_3$ material. Again the peak intensities observed at 270, 308, 525 and 725 cm^{-1} are assigned to the $A_1(TO_2)$, $E(TO_2)$, $A_1(TO_3)$ and $A_1(LO_3)$ of the P4mm phase of barium titatate.

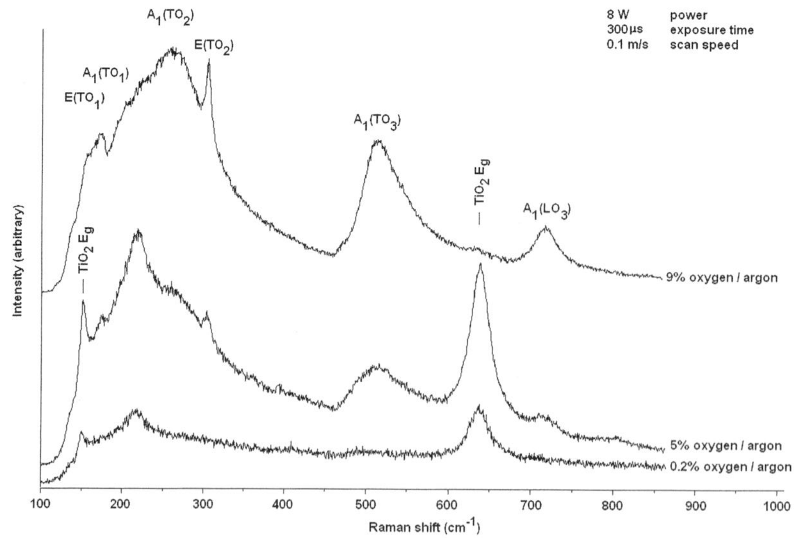

Fig. 3. Raman spectra of sintered $BaTiO_3$ as a function of oxygen partial pressure.

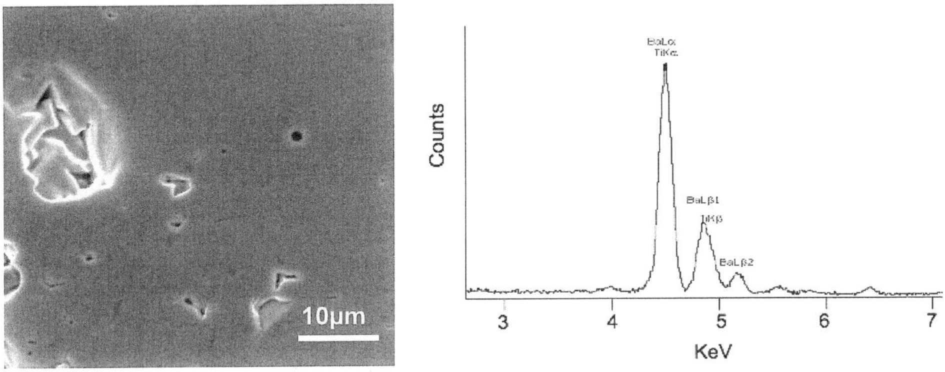

Fig. 4. (a) Scanning electron micrograph of a sintered sample $BaTiO_3$, (b) EDX spectrum of the area shown in 4(a).

As the partial pressure of oxygen is reduced, the low (144cm^{-1}) and high (639cm^{-1}) frequency E_g modes of anatase become evident in the sintered material as the extent of $BaTiO_3$ decomposition becomes more pronounced. Under the processing environment of 0.2% O_2 / Ar the whole sampled depth of the ceramic resembles TiO_2 without any evidence of the presence of barium titanate. This observation is consistent with the effect of laser power, but also suggests that a crucial interplay exists between the influence of laser power and oxygen ambient on the loss barium oxide from the perovskite titanate.

The microstructure of the sintered barium titanate was investigated using scanning electron microscopy. Fig. 4 shows an SEM image of a sintered sample of $BaTiO_3$ which shows no sign of decomposition in the associated Raman spectrum. The smooth surface morphology is indicative of extensive sintering, however regions of the inner structure of the sample are evident within the open pore structures. The internal microstructure shows a highly facetted structure with some penetrating porosity. An EDX spectrum of the area shown in Fig. 4(a) reveals that the composition of the SLS processed material retains the Ba, Ti and O (0.5KeV) constituents. These are in a similar ratio of intensities to those in the starting material. The Ba and Ti peaks at 4.465 keV and 4.508 keV respectively, can not be distinguished individually due to the inherent limited energy resolution of the EDX technique. This measure of elemental composition suggests that the correct combination of laser power and O_2/Ar atmosphere can suppress any undesirable laser induced modifications of the ceramic material.

The observations presented here indicate that only a limited range of laser powers is feasible for the formation of sintered barium titanate films at a given oxygen partial pressure. At high laser powers cracking of the sintered powders was observed. Fig. 5 shows optical micrographs of a sintered $BaTiO_3$ sample which exhibits cracking after treatment. The issue of cracking during the laser crystallization of a similar ceramic, namely $Ba_{0.7}Sr_{0.3}TiO_3$, using ArF and KrF excimer lasers has been investigated in detail [16]. The film thickness of the crystallized layer was observed to be an important parameter in minimizing crack formation. Thermal modelling was used to analyze the crystallisation process and it was concluded that the substrate temperature should not exceed 450 °C in order to eliminate cracking. In the present study, lower powers and higher O_2 / Ar ratios are necessary to avoid decomposition of the barium titanate but also to reduce cracking in the sintered layer .

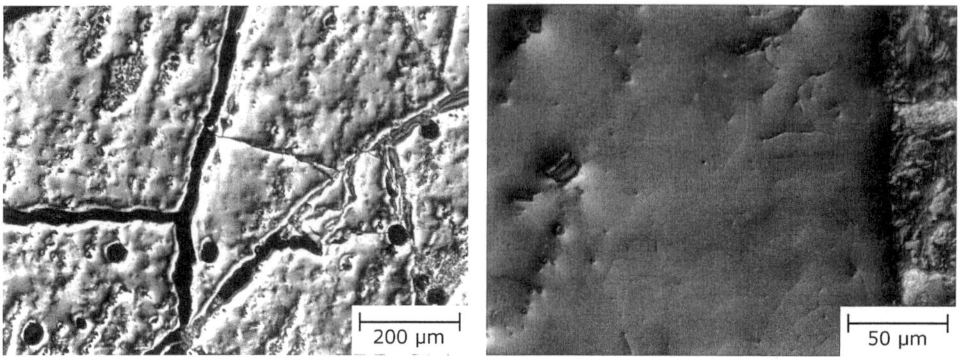

Fig. 5. Optical micrographs of a $BaTiO_3$ sample sintered at a laser power of 12W.

Conclusions

The results presented here indicate the potential for applying selective laser sintering in the manufacture of electroceramic films without the need for a mediating polymer matrix or other sintering agent. As SLS is a 'direct-write' process it has the capability of rapid-prototyping of complex structures without the need for lithographic masks and techniques. The selective laser sintering process has been investigated as a method for the fabrication of electroceramic parts from barium titanate powder. A range of laser power and oxygen partial pressures were identified in which sintered barium titanate layers could be formed. Insufficient concentrations of oxygen in the sintering ambient, resulted in the decomposition of the barium titanate to form an almost completely anatase-like surface as measured by Raman spectroscopy. A trade-off between laser power and oxygen content is observed if the sintering of $BaTiO_3$ is to be achieved.

Acknowledgements

The authors gratefully acknowledge the financial support of the Engineering and Physical Sciences Research Council under the project EP/D064856/1.

References

[1] Y. Zhang, L. Hao, M. M. Savalani, R. A. Harris, K. E. Tanner: J Biomedical Materials Research Part A, 86A (3) (2008) 607-616
[2] A. T. Clare, P. R. Chalker, S. Davies, C. J. Sutcliffe, S. Tsopanos: J. Mater. Sci., 43 (2008) 3197–3202
[3] D. Guo, L. T. Li, K. Cai, Z. L. Gui, C. W. Nan: J. Am. Ceram. Soc. 87 (1) (2004) 17-22
[4] B. Bihari, J. Kumar, G.T. Stauf, P.C. Van Buskirk, C.S. Hwang: J. Appl. Phys. 76 (1994) 1169.
[5] D.H. Kim, H.S. Kwok: Appl. Phys. Lett. 67 (1995) 1803.
[6] M.E. Lines, A.M. Glass: *Principles and Applications of Ferroelectrics and Related Materials*, Clarendon Press, Oxford, 1977.
[7] S.S. Thony, H.W. Lehmann, P. Gunter: Appl. Phys. Lett. 61 (1992) 373.
[8] C.H.J. Huang, T.A. Rabson: Opt. Lett. 18 (1993) 811.

[9] H.A. Lu, L.A. Wills, B.W. Wessels, W.P. Lin, T.G. Zhang, G.K. Wong, D.A. Neumayer, T.J. Marks: Appl. Phys. Lett. 62 (1993) 1314.
[10] D.M. Gill, B.A. Block, C.W. Conrad, B.W. Wessels, S.T. Ho: Appl. Phys. Lett. 69 (1996) 2968.
[11] N. Shu, A. Kumar, M.R. Alam, H.L. Chan, Q. You: Appl. Surf. Sci. 109/110 (1997) 366.
[12] K. Kaemmer, H. Huelz, B. Holzapfel, W. Haessler, L. Schultz: J. Phys. D: Appl. Phys. 30 (1997) 522.
[13] L. Beckers, J. Schubert, W. Zander, J. Ziesmann, A. Eckau, P. Leinenbach: J. Appl. Phys. 83 (1998) 3305.
[14] J.A. Sanjurjo, R.S. Katiyar, S.P.S. Porto: Phys. Rev. B 22 (1980) 2396.
[15] B.D. Begg, K.S. Finnie, E.R. Vance: J. Am. Ceram. Soc. 79 (1996) 2666.
[16] O. Baldus, R. Waser: Journal of the European Ceramic Society, 24, (10-11) (2004), 3013

Solid State Phenomena Vol. 154 (2009) pp 9-15
online at http://www.scientific.net

Finite Element Model of Polyelectrolyte Hydrogels Swelling – Comparison with Experiments

Robert A. Paxton & Ahmed M. Al-Jumaily
Institute of Biomedical Technologies
Auckland University of Technology
Auckland, New Zealand
ahmed.aljumaily@aut.ac.nz

Keywords: Finite element, electroactive, poly(acrylic acid)

Abstract: The preliminary results of gel swelling experiments are reported, and then compared to predictions made by a recently-developed finite element model (FEM). This model utilises energy transport between different energy domains, and is being used to simulate gel swelling dynamics. Initial experiments have revealed the model does capture the general behaviour of polymer hydrogel swelling dynamics and further improvements are necessary for better accuracy.

Introduction

For the past five years our group has been extensively studying the swelling dynamics of polymer hydrogels, for use as a novel type of changeable focal length lens [1]. It is anticipated that the "volume phase transition", inherent to all polymer hydrogels, can be utilised to create a lens with multiple focal distances. While hydrogel swelling can be triggered by a wide variety of stimuli, for the present purposes, electrical stimulation was seen as the optimum choice.

Much of our previous work has been experimentally based, and it was felt that a computational model was necessary to improve our underlying knowledge of the volume phase transition. To be of use, this model needed to be able to calculate the theoretical deformation of the hydrogel under different swelling conditions, including the effect of using different electrode locations/geometries, applied voltages and environmental conditions.

Due to the complexity of hydrogel swelling, analytical solutions were found to be unsuitable, and so numerical methods needed to be employed. Using MATLAB, a finite element model (FEM) was developed that could describe the swelling of a poly(acrylic acid) hydrogel under the influence of an applied electric field.

Model Summary

The developed finite element model [2] is based on the energy transfer between different energy domains, and consists of a series of electrical, chemical and mechanical components (Figure 1). It is able to be run without the need for a supercomputer, making it both versatile and portable. It is also highly customisable, and essential results at different stages of the model are available to the end user. The modular design of the model also permits each piece of the model to be improved individually, and then recombined back into the overall model.

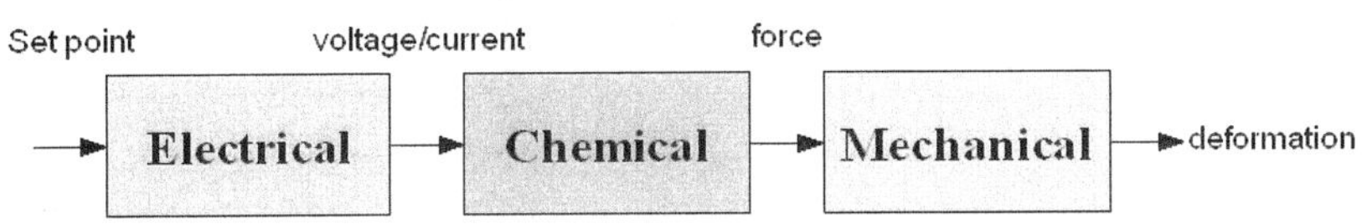

Figure 1: Component FEM model based on different energy domains

The Electrical component is the first part in the model and calculates the externally-applied electric field in the hydrogel region. This component incorporates Laplace's equation in a conductor to develop the electrical field distribution. Based on this, the Electrical module first assembles and then solves multiple, simultaneous instances of the Kirchoff current law (KCL) equations for each node in the system. This provides knowledge of the potential at each node, which the module can then use to calculate the externally-applied electric field.

The Chemical component is the second part of the model, and also forms its main engine. It is responsible for calculating the temporal and spatial changes in ion concentrations, resulting from the diffusion and migration of each species. The material flux in the hydrogel is calculated by iteratively solving the two-dimensional diffusion equation. Using these changes in ion concentrations, the model then calculates the osmotic pressure (force) on the hydrogel. Currently, these forces are analysed within the framework of the Flory-Rehner (F-R) theory [3], which assumes that the total osmotic pressure on the hydrogel is made up of three terms: the osmotic pressure due to ionic interactions (πion), the osmotic pressure due to polymer/solvent mixing (πmix) and the osmotic pressure due to the elasticity of the polymer chains (πelas). Although the F-R theory is generally regarded as only being accurate for ideal cases, most gel swelling models utilise it, and so it is used in this model without significant modification.

The Mechanical component is the final part of the model, and it provides the final output in terms of geometrical structure. The basis of this component is the standard equation of dynamic motion, using linear elasticity and constant mass. The Mechanical component receives information about the magnitude, direction and location of forces being applied to the hydrogel from the Chemical component. Using the material parameters of the hydrogel with these forces, the Mechanical component calculates the predicted nodal deformations resulting from these forces, which can then be displayed graphically if required (as shown in Figure 2).

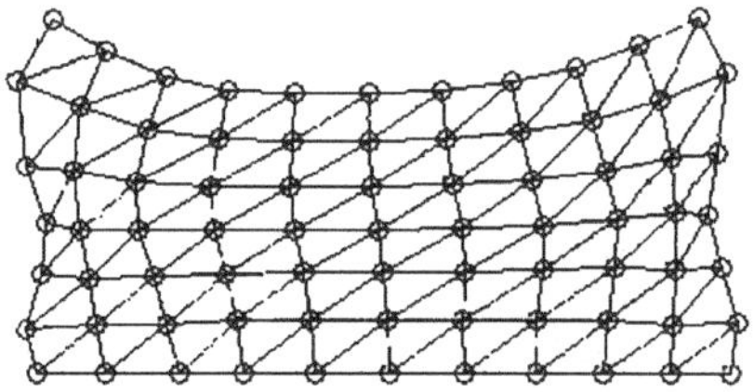

Figure 2: Theoretical deformation of hydrogel

In order to verify the FEM model, gel swelling experiments were conducted on poly(acrylic acid) hydrogel disks under the influence of an externally-applied DC source. The deformation of the upper surface of the gels was recorded, and then compared to the theoretically predicted deformation. It is worth mentioning that this is the preliminary stage of this model, and further improvement is underway.

Preparation of hydrogels

Poly (acrylic acid) (PAAC) hydrogels were created using thermally-triggered free radical polymerisation [4].

First, 4.8mL (70mmol) of acrylic acid (Alfa Aesar, 99%, used as received) was dissolved into 50mL of doubly-distilled deionised water (DDDI) to create a 1.277 mol L^{-1} solution. Although the AAC was inhibited with 200ppm 4-methoxyphenol (MEHQ), Cutié *et al.* [5] found that in isothermal persulfate-initiated polymerisations, MEHQ acted as an inhibitor and slowed, but did not prevent a polymerisation from occurring. For this reason, it was not deemed necessary to remove the MEHQ prior to polymerisation.

5.6mL (28mmol) of NaOH (Univar, analytical grade) was added to create a 40% neutralised pre-gel solution. 1.4mmol of N,N-methylene-*bis*-acrylamide (BIS) crosslinker was then added to make a lightly crosslinked hydrogel. In a previous work [1] it was shown that the optical transparency of PAAC hydrogels is inversely proportional to the amount of crosslinker used, and so it was important to use the smallest amount of crosslinker possible, whilst still retaining some of the structural properties of the hydrogel. The pre-gel solution was then placed under vacuum and stirred vigorously for 15 minutes to ensure that all components were thoroughly mixed. Although some gel-preparation methods pass nitrogen gas through the pre-gel solution to displace any free oxygen (which can act as an inhibitor) that may be present, this technique was found to have almost no effect on the produced hydrogel, and so was not used in this case.

After 15 minutes of stirring, the seal of the vacuum was broken and 0.04g potassium persulfate (KPS) (Scientific Supplies, AR grade, used as supplied) and 10μL of N,N,N',N'-tetramethylethylenediamine (TEMED) (Sigma, electrophoresis grade, used as supplied) were added to the pre-gel solution. KPS is a thermally-triggered polymerisation initiator, and is used to provide free radicals to the pre-gel solution. The use of TEMED is not essential for polymerisation to occur but it does act as an accelerator. The pre-gel solution was then again placed under vacuum and thoroughly stirred for a further 5 minutes.

The solution was then transferred to one or more glass petri dishes and placed into a vacuum oven (Lab-line, DuoVac 3610) at a temperature of 70°C and pressure of -44kPa until polymerisation was complete (typically 1.5 hours). Once completely polymerised, the hydrogel discs were removed from the oven and allowed to cool. When cooled, the PAAC hydrogel were thoroughly washed using DI water and allowed to equilibrate with the atmosphere for a period of 24 hours.

In order to conduct swelling experiments, smaller gel discs were cut from the Petri dishes using a customised glass cutting tool. The diameter of these discs was constrained by the cutting tool, while the height of the discs was controlled by varying the amount of pre-gel solution used during preparation. These discs were then used without any further modifications (Figure 3).

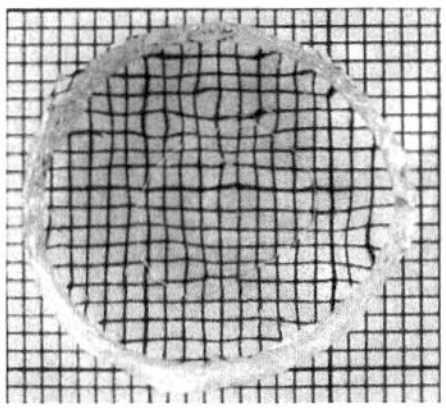

Figure 3: Smaller gel disks used for experimentation. Small square = 1mm

Gel Swelling Experiments

In order to investigate the swelling dynamics of PAAC, the smaller cuts discs were placed into 20ml of a 4 $molL^{-1}$ NaCl solution and subjected to external voltages of 5, 7.5 or 10V DC. This concentration of NaCl is significantly higher than that used by most other authors, but this is done deliberately to limit the overall swelling of the hydrogel disks.

Experiments were conducted using a ring anode (diameter 60mm, height 3mm) and a thin disc-shaped cathode (diameter 15mm), as shown in Figure 4. Both electrodes were made of high-purity electrical grade aluminium (Ling In Electronics Company Ltd), which was found experimentally to perform well in the current setup. The experiments were conducted in a large plastic container that was filled with a sufficient quantity of NaCl to completely cover the anode (approximately 5-10ml). To measure the deformation of the hydrogel disks, a digital video camera (Sony TRV25E) was used. The camera was placed adjacent to the gel/solvent system (Figure 5) and data recorded at a resolution of 720×625 pixels at 25fps. A square grid with 1mm spacing was placed behind the experiment during filming to enable the deformation of the upper surface to be measured (at 30s intervals). Using this deformation, the approximate volume change could then easily be calculated by assuming symmetry in the z-direction.

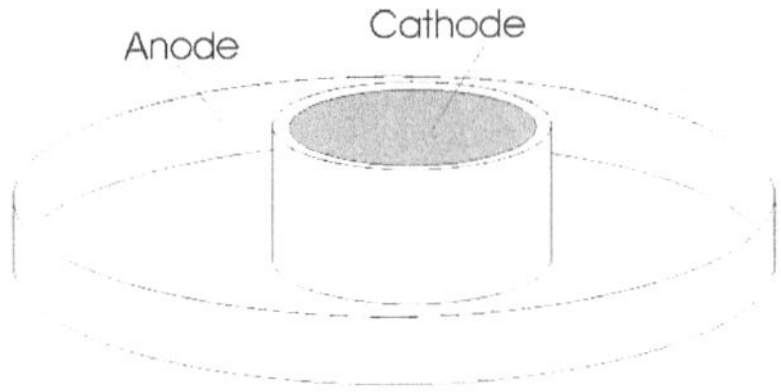

Figure 4: Disk and ring electrode geometry used in swelling experiments

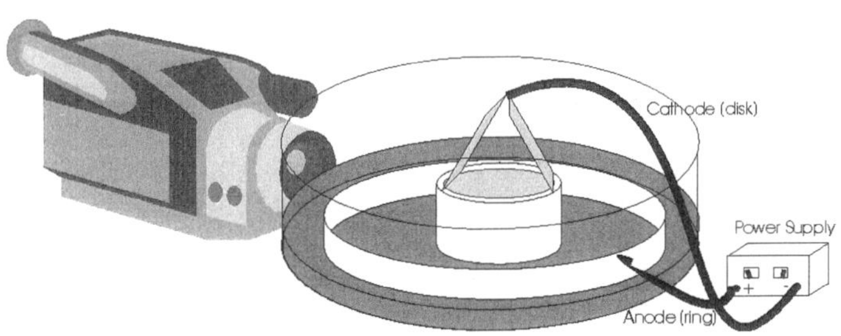

Figure 5: Camera setup used to record gel deformation

The deformation of the upper surface of the hydrogel disks was analysed using three points (upper left edge, centre and upper right edge) at 30s intervals (Figure 6) for a period of 5 minutes. By using knowledge of the diameter of the gel disks, the scale of the film was determined and the deformation of the upper surface then determined for various excitation voltages.

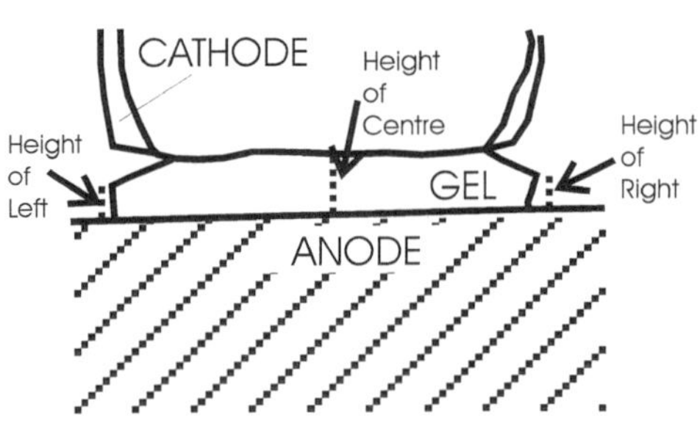

Figure 6: Data recorded for left, centre and right edges of hydrogel (front view)

Comparison of Results

Each experiment was performed several times, and the average deformation of the left, centre and right edges of the gel are shown in Figure 8. These results are then compared to the deformation predicted theoretically by the FEM model [2] (discussed previously).

Experimentally, the most noticeable difference between the upper left, right and centre edges of the samples is that the deformation of the centre is in the positive y-direction, whilst the deformation of the side nodes is in the negative y-direction. This was an unexpected and undesired behaviour of the system, and most probably stems from the electrode configuration used.

During the initial experimentation, it was found that when an electric field was applied, the hydrogel disks moved transversely, and consequently broke the electrical circuit (Figure 7). This movement was caused by an unbalanced force which is generated as the hydrogel disks swell. In order to prevent this "gel sliding" action, various electrode configurations were investigated that would constrain any transverse movement of the hydrogel while still permitting deformation in the vertical direction. The current experimental setup does permit the centre of the hydrogel disks to deform freely while simultaneously constraining any lateral movement. Methods of improving the electrical circuit are currently being investigated, and this will hopefully lead to a better method that will allow the hydrogel to move without limit, while still maintaining an electrical circuit.

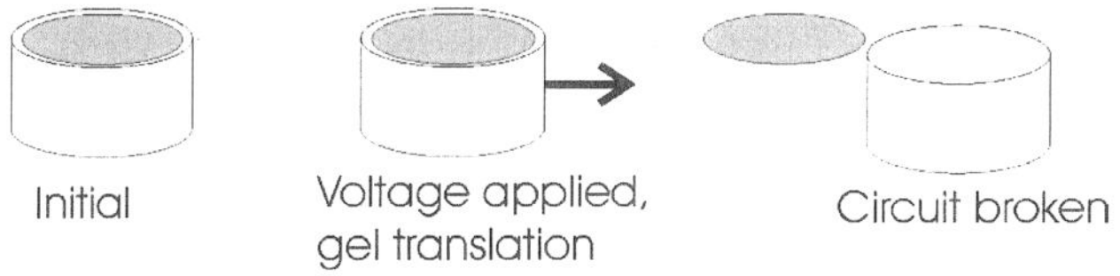

Figure 7: Translation of hydrogel under the influence of an applied electric field

The experimentally measured deformation of the upper centre of the hydrogel was linear in shape, and the rate of deformation (and therefore overall deformation) increased with increasing voltage. This is contrary to the results obtained by Salehpoor *et. al* [6], who found that the overall deformation was independent of the applied voltage, and only the swelling rate was affected. It is possible that this work would obtain similar results to that of Salehpoor *et. al*, if the swelling was measured over a longer period of time, and this is being investigated for future work. It is clearly indicated that the experimental data follows a linear trend, while the theoretical data follows an exponential trend. The magnitude of the experimentally measured deformation is also higher than that predicted by the model.

The experimentally measured deformation of the upper left and right edges of the hydrogel appears to follow a decreasing trend (as discussed previously) and also seem to indicate that the swelling rate decreases with increasing applied voltage. The magnitude of upper left and right deformation is larger than the deformation of the centre, when the experimental data is again compared to the theoretical predictions (Figure 8). The theoretical swelling of the right edge is also noticeably smaller than the swelling of the left edge, although the exact reasons for this are currently being investigated.

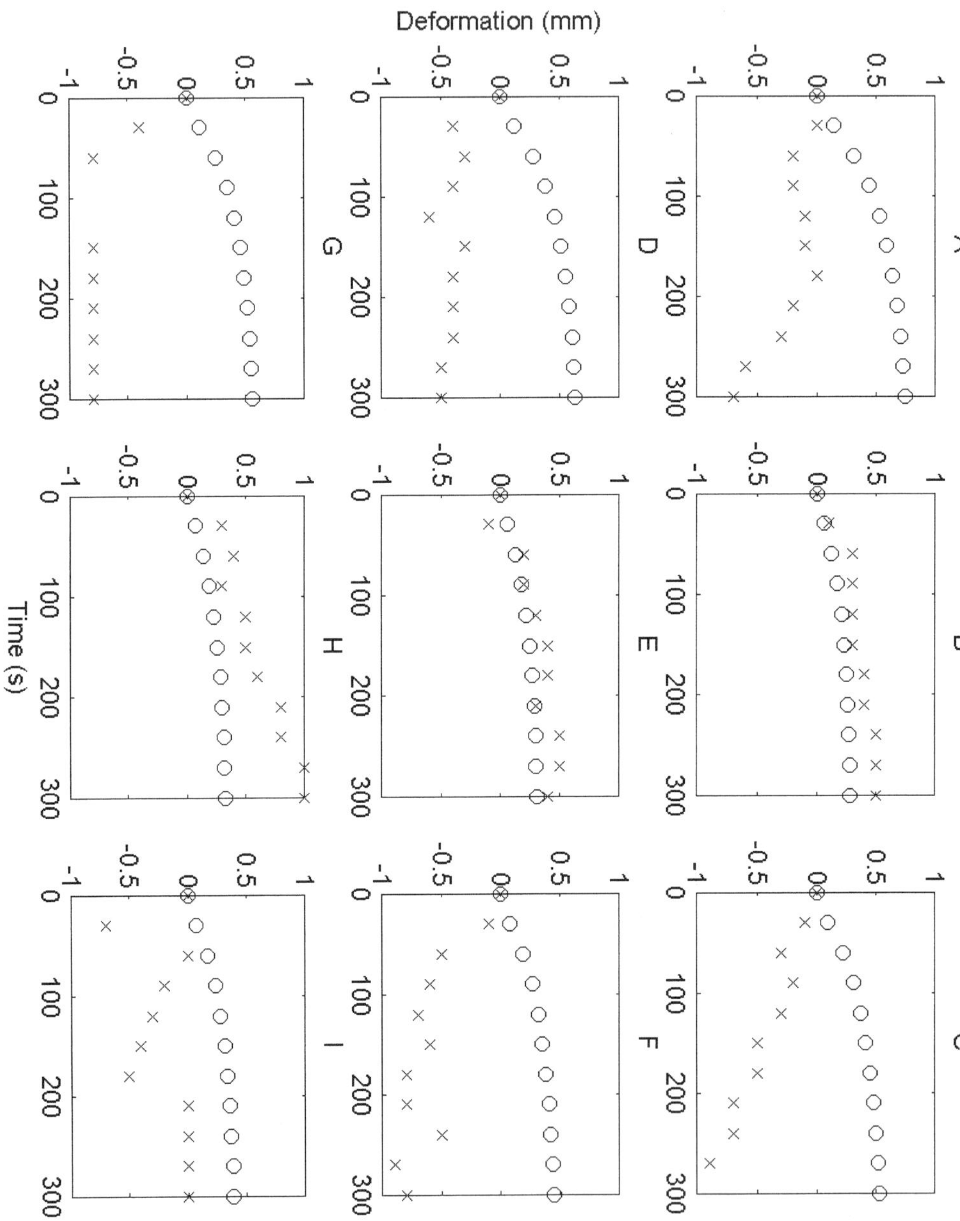

Figure 8: Theoretical (o) and experimental (x) deformation for poly(acrylic acid) hydrogels [A,B,C = deformation of left, centre & right for 5V; D,E,F = deformation of left, centre & right for 7.5V; G,H,I = deformation of left, centre & right for 10V]

Discussion

Comparing the results of the model with those of the experiments indicates that the model does predict the general behaviour of the gel swelling process. However, further improvements are necessary for better accuracy. Possible improvements include the inclusion of a "force feedback" loop which would undoubtedly change the trend of the theoretical swelling results. Using the results obtained from this set of experiments, this and other modifications and improvements are constantly being made to the model. One interesting prediction made by the model was that the swelling rate of the centre of the hydrogel increases with increasing voltage. Similar behaviour was expected for the left and right edges. However, the swelling rate of the centre of the left and right edges of the hydrogel decreases with increasing voltage – an effect also observed experimentally.

The experimental setup also requires improvement, to allow more accurate deformation measurements to be made. The experimental lighting conditions and electrode geometry also need to be investigated in order to find a system that does not constrain the hydrogel, as discussed previously.

Conclusion

This paper reports on the first set of experiments used to validate an FEM model to describe the swelling of polymer hydrogels. While the results indicate that much work still remains to be done on the model, the preliminary results are better than initially anticipated. The model does correctly identify the relationship between the overall deformation and the applied voltage. While the current model predicts that the swelling rate is independent of applied voltage, this most probably results from the lack of a feedback loop in the model. This issue is currently being addressed, and will be reported in later publications.

References

1. Paxton RA. M.E. Thesis, Department of Mechanical Engineering. Auckland: Auckland University of Technology; 2002.
2. Paxton RA. PhD Thesis, Department of Mechanical Engineering. Auckland: Auckland University of Technology; 2006.
3. Durmaz S, Okay O. Polymer 2000; 41(3693-3704.
4. Paxton R, Al-Jumaily AM, Easteal AJ. Polymer Testing 2003; 22(371-374.
5. Cutié SS, Henton DE, Powell C, Reim RE, Smith PB, Staples TL. Journal of Applied Polymer Science 1997; 64(3): 577-589.
6. Salehpoor K, Shahinpoor M, Mojarrad M. Proceedings of SPIE 1996; 2716: 36-45

Solid State Phenomena Vol. 154 (2009) pp 17-22
online at http://www.scientific.net

Kinetics of swelling and drug release from PNIPAAm/alginate stimuli responsive hydrogels

Raluca Petronela Dumitriu[1,a], Ana Maria Oprea[1,b], Cornelia Vasile[1,c]

[1]„P. Poni" Institute of Macromolecular Chemistry, Physical Chemistry of Polymers Department, 41 A Gr. Ghica Voda Alley, 700487, Iasi, Romania,

[a]e-mail: rdumi@icmpp.ro, [b]e-mail: aoprea@icmpp.ro, [c]e-mail: cvasile@icmpp.ro

Keywords: hydrogels, stimuli-responsive, swelling behavior, drug delivery, kinetics systems

Abstract. Stimuli responsive hydrogels are very attractive for applications in sustained and/or targeted drug delivery systems. As the release of drugs is related to the swelling behavior of hydrogels, the swelling kinetic studies become of great importance to appreciate the release kinetics from hydrogel matrices. Hydrogels with high performance properties have been prepared from N-isopropylacryl amide (NIPAAm) and sodium alginate, crosslinked with N,N`-methylene-bis-(acrylamide) (MBAAm). This study is focused on the investigation of swelling and drug release kinetics, coupled by morphological studies. The kinetic parameters of the swelling at different temperatures for hydrogels samples have been evaluated and confirmed their temperature-responsive behavior. The swelling rate constant (k_{sw}) decreases of with increasing temperature and slight increases with the alginate content in the samples. The drug release kinetic study from the prepared hydrogel matrices was performed in twice-distilled water and ethanol for bioactive agents as vanillin and ketoprofen, respectively. An increase of alginate content results in a slower rate and smaller percentage of vanillin and ketoprofen released. It has been established that the ketoprofen occurs according with case II of transport and vanillin release behavior occurs by an anomalous transport mechanism. The values of the release rate constant (k_r) decreased by increasing swelling degree in case of 75/25 NIPAAm/alginate hydrogels and decreased also by increasing content of alginate in hydrogels with various compositions. Morphological studies performed by environmental scanning electron microscopy (ESEM) evidenced a relaxed network at high relative humidity, which explain both swelling and release profiles.

Introduction

Over the last years, the biomedical applications of hydrogel type materials has known an impressive development, which created the need to develop new methods for their synthesis and controlled delivery of active principles.

Smart hydrogels were investigated as drug delivery systems with ability to control the release of drugs in response to changes in the surrounding environment, such as ionic strength, pH, light, electric field, irradiation, or temperature. [1,2,3]

Among the temperature-sensitive hydrogels reported to date, poly (*N*-isopropylacrylamide) (PNIPAAm) and its copolymers have been widely used for cell separation as well as for pharmaceutical and tissue engineering applications because of their unique thermo-responsive properties [4,5,6]. PNIPAAm undergoes a reversible temperature-sensitive coil-to-globule transition in aqueous solutions at approximately 32 ^{0}C (its lower critical solution temperature, LCST), which lies between room temperature and physiological temperature (37 ^{0}C). The LCST of PNIPAAm can be increased by copolymerization with hydrophilic monomers or decreased by using hydrophobic monomers [7,8,9]

Alginate, a polysaccharide extracted from brown algae has been used as a hydrophilic component of hydrogels for many biomedical applications. The ability of alginate sodium salt to rapidly form viscous solutions and gels in contact with aqueous media has been exploited by the pharmaceutical industry, being applied as a carrier (hydrophilic matrix) for controlled release of oral dosage forms. Matrices incorporating alginate salts or a combination of alginate with other polymers have been employed to successfully prolong release of many drugs [10,11,12]

The swelling phenomena and its kinetics have been long studied. Most of the past studies were focused on the swelling equilibrium. A better understanding of the swelling/shrinking kinetics of the hydrogels is vitally important for their potential applications. The attention granted to the studies on the swelling behavior of hydrogels may be justified by the fact that not only the delivery of drugs is accomplished by the swelling characteristics of hydrogels (i.e. pH-sensitive hydrogels) [13], but the release of drugs is also determined by their swelling behavior [14].

The goal of this paper is to investigate both the kinetics of swelling and the ketoprofen or vanillin release from N-isopropylacryl amide (NIPAAm)/alginate (Alg) hydrogel matrices.

Experimental

The monomer N-isopropylacryl amide (Aldrich) 97% (NIPAAm), the polysaccharide alginic acid extracted from brown algae (Fluka), M_w = 48 000 - 186 000 with reduced viscosity at 25^{0}C of 2.41 ml·g^{-1}, c = 0.2 g/dL (Alg) and the crosslinking agent N,N'-methylene bis-acrylamide (Fluka) were all used as received. The twice-distilled water was employed for all swelling experiments. The hydrogel samples were prepared in various mixing ratios: 99/1, 95/5, 90/10, 85/15, 80/20, 75/25 NIPAAm/Alg, purified by five repeated washings with twice-distilled water and then freeze - dried.

Ketoprofen is a classical nonsteroidal anti-inflammatory (first generation), analgesic and antipyretic drug efficient in the treatment of the rheumatoid arthritis and osteoarthritis, spondylitis, as well as soft tissue injuries and also to reduce fever. It is a water insoluble substance but it dissolves in solvents like ethanol, methylene chloride, dimethylsulphoxide (DMSO) and dimethylformamide (DMF).

Vanillin (3-methoxy-4-hydroxybenzaldehyde) possesses antifungal and antibacterial activity to a very high extent, with very broad spectrum coverage against dermatophytes, *Candida*, other yeast fungi, and mold fungi including *Trichophyton interdigital* and *Trichophyton rubrum*, the athlete's foot molds, as well as Gram-positive and Gram-negative bacteria.

The *kinetics of the swelling* was carried out by weight measurements performed at different temperatures, in the range 20 – 40 ^{0}C in twice-distilled water. The equilibrium swelling degree was calculated according to the Eq. (1).

$$Q_{eq} = (W_{eq} - W_d)/W_d \times 100(\%) . \quad (1)$$

where W_{eq} is the weight of the swollen sample when thermodynamic equilibrium was reached and W_d is the dry weight of the sample.

The *drug loading* of the hydrogel matrices was carried out by mixing the drugs used, ketoprofen and vanillin respectively, with dried hydrogel in powder form and then a certain quantity of the appropriate solvent (ethanol for ketoprofen and twice-distilled water for vanillin) was added and left to swell at room temperature, while the drugs penetrate and/or attached into matrices. The drug-loaded samples were freeze-dried using a Labconco FreeZone device. During the *drug release* study, at predetermined time intervals, 1 ml sample was withdrawn from the release medium and concentration of ketoprofen at 254 nm and of vanillin at 228 nm in the release medium were determined using a UV-VIS spectrophotometer HP 8450A.

To determine the kinetics of solvent diffusion into the hydrogels the following equation (Eq. 2) was used:

$$F_t = \frac{W_t}{W_{eq}} = k_{sw} t^{n_{sw}} \quad . \tag{2}$$

where W_t and W_{eq} represent the amount of water absorbed by the hydrogel at time t and at equilibrium respectively, k_{sw} is the swelling constant characteristic of the system and n_{sw} is the power law diffusion exponent which takes into account the type of solvent transport. Eq. 2 applies to initial states of swelling and linearity is observed when log F_t as a function of log t is represented.

In order to elucidate the kinetics of drug release, the data were further analyzed using the equation proposed by Korsmeyer and Peppas (Eq.3):

$$M_t / M_\infty = k_r t^{n_r} \quad . \tag{3}$$

where M_t/M_α represents the fraction of the drug released at time t, k_r is a constant incorporating characteristics of the macromolecular network system and n_r is the diffusion exponent, which is indicative of the release mechanism.

In the equations above a value of $n_{sw}/n_r = 0.5$ indicates a Fickian diffusion mechanism of solvent/drug in hydrogels, while a value $0.5 < n_{sw}/n_r < 1$ indicates an anomalous or non-Fickian behavior. When $n_r = 1$ a case II transport mechanism is involved while $n_r > 1$ indicates a special case II transport mechanism. [15,16]

The *morphological aspect* of the samples was investigated by means of a FEI Quanta 600 FEG Environmental Scanning Electron Microscope and the images were taken at different degrees of relative humidity (RH %).

Results and discussion

Swelling kinetic studies

In Table 1 the equilibrium swelling degree in water and the kinetic parameters of swelling are given for NIPAAm/Alg hydrogel samples with various compositions.

Table 1. Kinetic parameters of swelling for NIPAAm/Alg hydrogels

Hydrogel		T=20^{0}C			T=35^{0}C	
	Q_{eq} [%]	k_{sw} [min^{-n}]	n_{sw}	Q_{eq} [%]	k_{sw} [min^{-n}]	n_{sw}
99 NIPAAm /1Alg	2385	0.1	0.38	204	0.03	0.5
95 NIPAAm/5 Alg	2436	0.07	0.42	294	0.1	0.32
90 NIPAAm/10 Alg	3199	0.066	0.54	569	0.14	0.31
85 NIPAAm /15 Alg	2800	0.11	0.3	532	-	-
80 NIPAAm /20 Alg	3363	0.16	0.3	423	-	-
75 NIPAAm/25 Alg	3634	0.09	0.41	843	-	-

The values obtained for the equilibrium swelling degree (Q_{eq}) at different temperatures for the hydrogel samples with different compositions (Table 1) showed the decreased swelling capacity at temperatures higher than LCST of PNIPAAm (~ 32 ^{0}C) and the increased Q_{eq} values for hydrogels with higher content of alginate. This behavior can be explained due to the hydrophilic character of alginate.

In Figs. 1 and 2 the kinetic constant k_{sw} is represented as a function of temperature and alginate content in the samples. A decrease of k_{sw} values with increasing temperature is observed, excepting the case of temperature 35 ^{0}C, when the gels collapsed and their swelling behavior is influenced.

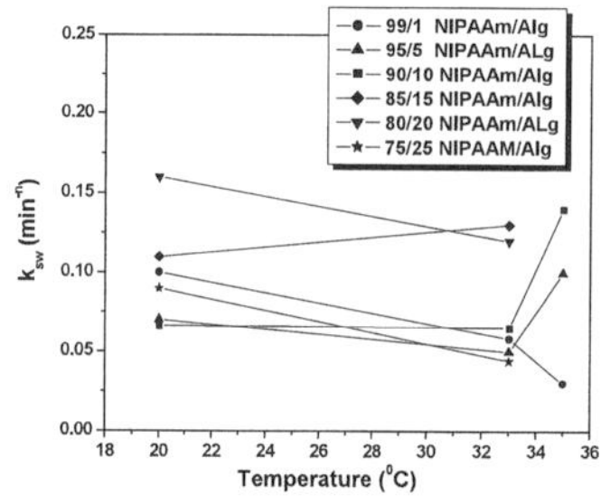

Fig. 1. Swelling constant as a function of temperature for NIPAAm/Alg hydrogels of various compositions

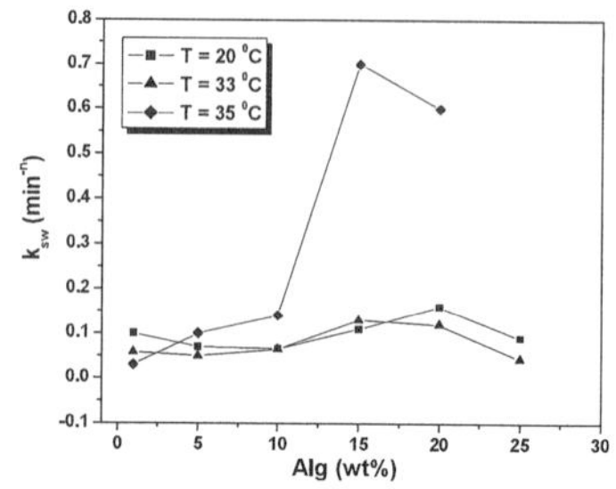

Fig. 2. Swelling constant as a function of alginate content in the NIPAAm/Alg hydrogels at different temperatures

The composition of hydrogels influences the swelling kinetics of the samples (Fig. 2) by increasing k_{SW} values with increasing alginate content. The change is observed especially at 35 ^{0}C and in the case of hydrogels with high Alg content above the LCST of PNIPAAm, when the hydrophilic character of alginate influenced the swelling behavior.

Kinetics of drug release

The influence of the swelling degree measured in ethanol on the release rate constant of *ketoprofen* from 75/25 NIPAAm/Alg hydrogel sample is represented in Fig. 3.

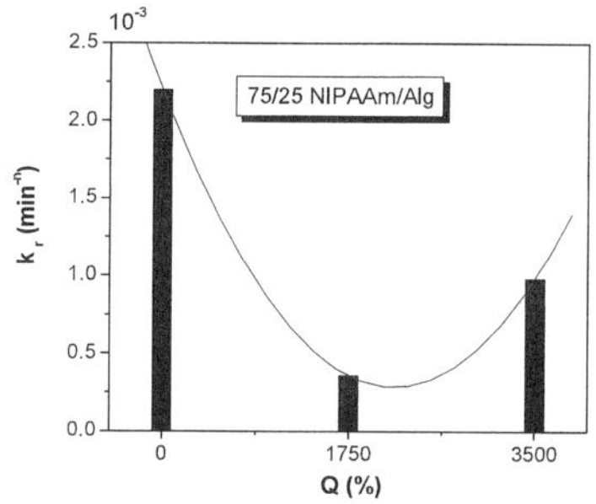

Fig. 3. Release rate constant dependence on the swelling degree for 75/25 NIPAAm/Alg hydrogel in case of ketoprofen release in ethanol

The results obtained showed differences between release rate constant values due to the influence of swelling on drug loading/release process, so in case of wetting the intermolecular spaces became larger and the drug penetrates and attaches into matrices determining a decrease of ketoprofen release rate constant values. Hydrogels with a 75/25 NIPAAm/Alg ratio, present in their dry state a four time bigger release rate constant value comparative to those obtained in case of the swelling degree between 1750 and 3500 %. The value of diffusion exponent $n_r \approx 1$ indicates a case II transport mechanism, determined by polymer relaxation as a consequence of swelling process in ethanol.

The kinetic parameters for *vanillin* release in twice-distilled water from NIPAAm/Alg hydrogels with various compositions are presented in Table 2.

Table 2. The kinetic parameters of the vanillin release from NIPAAm/Alg hydrogels

Hydrogels	n_r	k_r [min^{-n}]
95/5 NIPAAm/Alg	0.97	$1{,}1 \cdot 10^{-3} min^{-0.97}$
85/15 NIPAAm/Alg	0.70	$3{,}2 \cdot 10^{-3} min^{-0.70}$
75/25 NIPAAm/Alg	0.79	$9{,}4 \cdot 10^{-4} min^{-0.79}$

The alginate content in NIPAAm/Alg hydrogels and release rate constant values are connected, so the increase of alginate content lead to a decrease of release rate constant values and diffusion exponent (n_r) from $1{,}1 \cdot 10^{-3} min^{-0.99}$ for hydrogels with 95/5 composition to $9{,}4 \cdot 10^{-4} min^{-0.79}$ in hydrogels with 75/25 composition.

The values of the diffusion exponent n_r in that case, are 0.97 for the 95/5 composition, 0.70 for the hydrogel with 85/15 composition and 0.79 for the hydrogel with 75/25 composition, indicating an anomalous transport, which appeared by coupling Fickian diffusion with the relaxation of the hydrogel network.

Environmental scanning electron microscopy (ESEM)

The morphological study of the samples performed by ESEM evidenced a porous structure with honeycomb-like aspect, which varies for samples having different composition and modifies by expanding as hydrogels are swelling at increasing degree of humidity (Fig. 4).

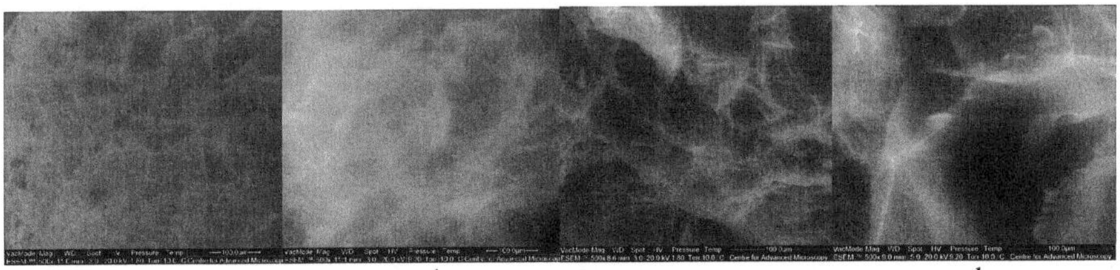

a b c d

Fig. 4. ESEM micrographs of the hydrogel samples 99/1 NIPAAm/Alg (a) RH=20 and (b) RH=100; 75/25 NIPAAm/Alg (c) RH=20 and (d) RH=100

It can be noticed that cavities dimensions are increased at higher relative humidity RH (%) for all the samples: the pores became larger or their walls are breaking, making the structure more accessible to water vapor sorption. Thus, the morphology of the samples is changing with increasing RH from a compact/dense structure to an expanded one. Samples with higher alginate content (75/25) have larger cavities with porous walls also, which favors water diffusion in the hydrogels, explaining their enhanced swelling capacity and the values obtained by swelling kinetic measurements and their behavior during drug loading also.

Conclusions

The profiles of the swelling and drug release showed a temperature-responsive behavior of the samples depending on their composition. The water sorption ability is increased with the alginate content and diminished at higher temperatures. The swelling exponent values $n_{sw} < 0.5$ indicated a non-Fickian character of water diffusion into the hydrogels and the swelling constant (k_{sw}) values are slightly decreased with increasing temperature due to collapse of the hydrogels above the LCST, excepting for the compositions with increased alginate content, which show a particular behavior.

The total relaxation of hydrogels network observed by ESEM at higher RH obtained at equilibrium swelling provides a better drug entrapment inside matrices and its slow release, the system being much suitable for a sustained drug release.

References

[1] B. Jeong, S.W. Kim, Y.H. Bae: *Adv. Drug Deliv. Rev* Vol. 54 (2002), p. 37-51

[2] I.Y. Galaev, B. Mattiasson: *Trends Biotechnol* Vol. 17 (1999), p. 335-340

[3] E. Ruel-Gariepy, J.C. Leroux: *Eur. J. Pharm. Biopharm* Vol. 58 (2004), p. 409-426

[4] C.A. Kavanagh, Y.A. Rochev, W.M. Gallagher, K.A. Dawson, A.K. Keenan: *Pharmacoll Ther.*, Vol. 102 (2004), p. 1-15

[5] E. Piskin, S. Dincer, M. Turk: *J. Biomater. Sci., Polym. Ed* Vol. 15 (2004), p. 1181-1202

[6] M. Hirose, M. Yamato, O.H. Kwon, M. Harimoto, A. Kushida, T. Shimizu, A. Kikuchi, T. Okano: *Yonsei Med. J* Vol. 41 (2000), p. 803-813

[7] D. Schmaljohann, J. Oswald, B. Jorgensen, M. Nitschke, D. Beyerlein, C. Werner: *Biomacromolecules* Vol. 4 (2003), p. 1733-1739

[8] D. Neradovic, O. Soga, C.F. Van Nostrum, W.E. Hennink: *Biomaterials* Vol. 25 (2004), p. 2409-2418

[9] R. M. K. Ramanan, P. Chellamuthu, L. Tang, K. T. Nguyen: *Biotechnol. Prog* Vol. 22 (2006), p. 118-125

[10] M. Efentakis, G. Buckton: *Pharm. Dev. Technol* Vol. 7 (2002), p. 69–77

[11] C.V. Liew, L.W. Chan, A.L. Ching, P.W.S. Heng: *Int. J. Pharm* Vol. 309 (2006), p. 25–37

[12] P. Sriamornsak, S. Sungthongjeen : *AAPS PharmSci- Tech* (2008), accepted for publication

[13] Y.H. Yin, Y.J. Yang, H.B. Xu: *Journal of Polymer Science, Part B: Polymer Physics* Vol. 39 (2001), p. 3128–3137

[14] Y.H. Yin, Y.J. Yang, H.B. Xu: *Journal of Applied Polymer Science* Vol. 83 (2002), p. 2835–2842

[15] I. Katime, R. Novoa, F. Zuluaga: *European Polymer Journal* Vol. 37 (2001), p. 1465-1471

[16] R.W. Korsmeyer, N.A. Peppas: *J. Control. Rel* Vol. 1 (1984), p. 89

Solid State Phenomena Vol. 154 (2009) pp 23-27
online at http://www.scientific.net

Differential magnetoelastic compressive force sensor utilizing two amorphous alloy ring cores

Adam Bieńkowski[1,a], Roman Szewczyk[1,b], Jacek Salach[1], Roman Kolano[2,c]

[1] Institute of Metrology and Biomedical Engineering, Warsaw University of Technology, sw. Andrzeja Boboli 8, Warszawa 05-525, Poland

[2]Institute of Nonferrous Metals, Sowińskiego 5, Gliwice 44-100, Poland

[a]bienko@mchtr.pw.edu.pl, [b]szewczyk@mchtr.pw.edu.pl, [c]romank@imn.gliwice.pl

Keywords: amorphous alloys, magnetoelastic effect, force sensors.

Abstract. Paper presents the results of investigation on functional characteristics of magnetoelastic compressive force sensor utilizing two $Fe_{81}Si_4B_{15}$ amorphous alloy ring-shaped cores. Uniform distribution of stresses in one of two cores was achieved owing to special non-magnetic backings. Signal from sensing coils of cores was connected to differential amplifier, whereas sine wave voltage signal was applied to magnetizing circuit. High stress sensitivity of developed sensor was indicated together with expected reduction of temperature sensitivity of magnetoelastic sensor. These results confirm, that differential configuration of magnetoelastic sensor is suitable for practical application in sensors development.

Introduction

Magnetoelastic Villari effect is connected with the changes of total free energy due to appearance of magnetoelastic energy in the material subjected to stresses from external forces [1]. This effect may be observed as changes of flux density B achieved in the magnetic core, for given magnetizing field H_m, when it is subjected to mechanical stresses σ [2]. From the practical point of view, the magnetoelastic properties of the material may be described by both $B(H)_\sigma$ □and $B(\sigma)_{Hm}$ characteristics, which represent the influence of stresses on magnetic hysteresis loops of the material, for both quasi-static and dynamic magnetization [3].

Magnetoelastic effects in the magnetic sample may be generated by external forces, resulting from hydrostatic pressure [4], compressive [5] or tensile forces [6], or torque [7]. The influence of compressive stresses σ on the magnetic characteristics is very interesting from practical point of view, because of the high stress sensitivity of amorphous alloys [8]. On the other hand, the magnetic properties of these materials also change, as the function of temperature of the core. To minimize the influence of this effect, unstressed sample should be used for compensation. In this paper, construction of such differential magnetoelastic sensor is presented.

Magnetomechanical sensing device

In the case of development of magnetoelastic stress sensors based on the ring cores made of annealed amorphous alloys, the method of applying stresses to the sensing element has to create the possibility of achievement of uniform stresses distribution. Due to brittleness of amorphous alloys in annealed state [9], non-uniformity of stresses may lead to creation of highly stressed areas, and damage of the sensing element. On the other hand, utilization of amorphous alloy cores in as-quenched state (when such cores don't exhibit significant brittleness) leads to decrease of stress sensitivity of the core, due to residual stresses generated in the core during the rapid quenching process [9].

To overcome these problems, the idea of novel method of the applying the compressive stress to the ring-shaped sensing element was developed. In this method the compressive force F is applied to the ring-shaped cores in direction of its axis. As a result, compressive stresses in the core are perpendicular to the magnetizing field H [4]. Moreover, in such case a uniform, compressive stress σ can be reached on the whole length of the magnetic circuit of the sensing element. In addition, the commercially available, annealed ring shaped cores can be used as the magnetoelastic stress and force sensors [8].

Device utilizing this idea is presented in figure 1. Compressive force F is applied to device by hydraulic press via the base backings (1). Between nonmagnetic backing (2) and the ring core (3) a special elastic spacer is placed, to guarantee uniform distribution of the stresses σ in the core. Magnetizing and measuring windings are placed in special grooved races (2a) in backings (2) [10]. As a result, windings are not subjected to stresses, and not deformed by application of force F.

From technical point of view, in presented configuration stresses up to 100 MPa can be reached in tested samples. On the other hand, the value of stresses is limited by the sample stress strength.

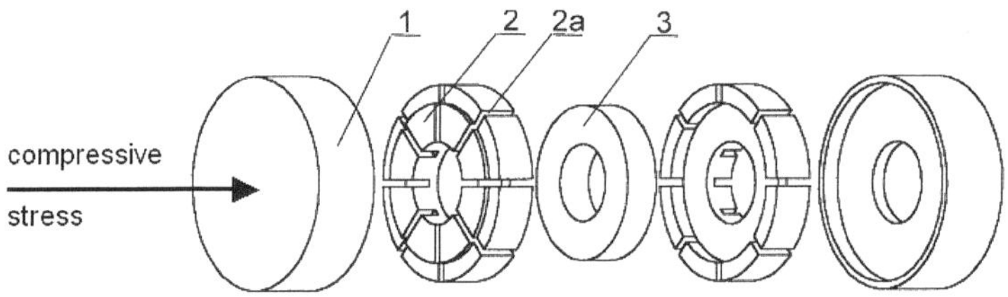

Fig. 1. Device for application of the uniform compressive stress to the ring-shaped sensing element [10]: 1 – base backings, 2 – nonmagnetic cylindrical backings, 2a – grooves or the winding, 3 – core under investigation

Two cores made of $Fe_{81}Si_4B_{14}$ rapidly quenched amorphous alloy ribbon were used for the development of differential magnetoelastic sensor. The amorphous ribbon thickness was 25 μm. The outside diameter of cores was 32 mm, inside diameter was 25 mm, whereas height was 10 mm. Both samples were annealed in 300 °C for one hour in magnetic field H_a = 150 kA/m, perpendicular to the ribbon's direction. That value of magnetic field H_a, applied during the annealing, induced the perpendicular anisotropy in samples, which increases the stress sensitivity of amorphous alloy.

Electronic signal processing

A schematic block diagram of differential magnetoelastic sensor's electronic unit is presented in figure 2.

Sine-wave generator is based on the Wien bridge, and generates signal with frequency of about 400 Hz. Amplitude of this signal is critical from the point of view of repeatability of the sensor. Due For that reason, it is being stabilized by a specialized proportional-integral amplitude controller, connected with precise voltage source based on integrated circuit. Then, the sine-wave signal from voltage-current converter drives magnetizing windings of both magnetoelastic cores connected in series. As a result, the signals driving both cores are in phase.

Signal from sensing windings of both the magnetoelastic core subjected to stresses, as well as the reference magnetoelastic core, is pre-amplified. Next, the difference of these two signals is amplified in differential amplifier. Then, the value of mean voltage of differential signal is rectified, and through the zero-shifter circuit, followed by active low-pass filter, drives an output to signal amplifier.

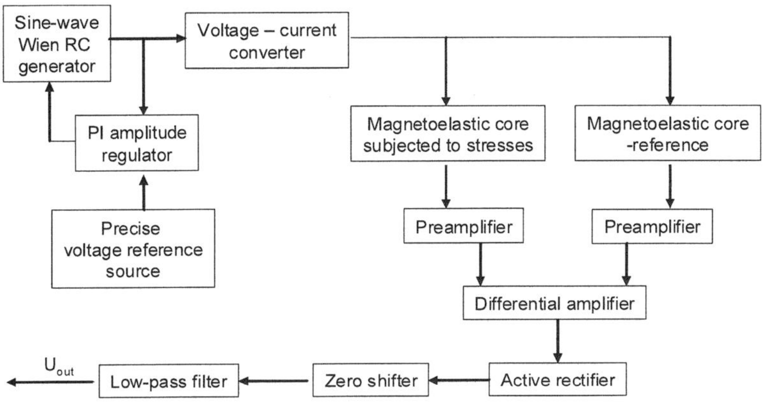

Fig. 2. Schematic block diagram of the electronic circuit of the differential magnetoelastic sensor

Results of differential magnetoelastic sensor testing

Figure 3 presents the results of testing the influence of compressive stress σ on the shape of the quasi-static hysteresis loop $B(H)_\sigma$ of magnetoelastic sensing, annealed in 300 °C for one hour, in magnetic field H_a = 150 kA/m, perpendicular to the ribbon's direction.

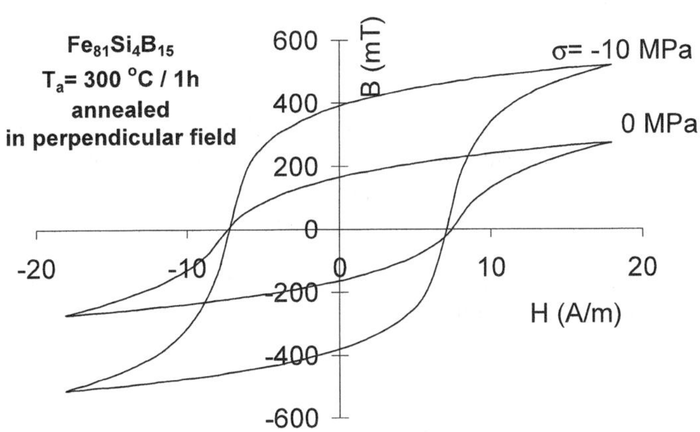

Fig. 3. $B(H)_\sigma$ characteristics of $Fe_{81}Si_4B_{14}$ amorphous alloy with perpendicular anisotropy

Under the compressive stresses σ up to 10 MPa the value of flux density B in the sample increases by more than 100%, whereas coercive force H_c of sample is nearly constant. These results confirm the high stress sensitivity of core used for development of differential magnetoelastic sensor.

In figure 4 the output characteristic of the developed differential magnetoelastic sensor is presented. Unfortunately, this characteristic is non-monotonous and reaches minimum for compressive force at about 500 N. This minimum is connected with the occurrence of higher harmonics in changes of magnetoelastic core's output signal, during the dynamic magnetization.

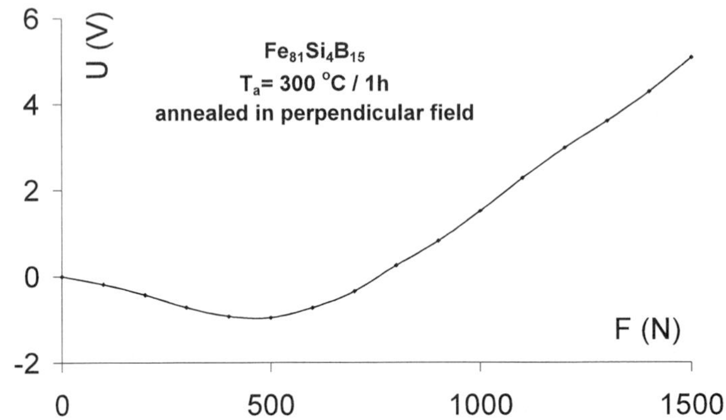

Fig. 4. Output characteristic of developed differential magnetoelastic sensor

Due to the occurrence of minimum on the sensor characteristic, the developed differential magnetoelastic sensor may work being pre-stressed by force of the value higher than 500 N. In such case it exhibits high stress sensitivity (about 4 mV/N) and quite wide measuring range. To verify the functional properties of the developed sensor, an estimation of standard deviation of repeatability was carried out. Standard deviation of repeatability is lower than 16 mV, what is sufficient for practical applications.

Conclusion

Presented results confirm the usability of proposed differential signal processing unit. The developed differential magnetoelastic sensor exhibits high stress sensitivity, as well as repeatability sufficient for practical applications. High repeatability confirms also the correct operation of specialized proportional-integral amplitude controller in Wien bridge generator. On the other hand, due to the occurrence of the minimum at sensor's output characteristic, it requires pre-stressing. This effect may be a barrier in practical application of the developed sensor. As a result, the parameters of thermo-magnetic treatment of magnetoelastic core should be optimized, to achieve the monotonous characteristics of magnetoelastic sensor operating in differential mode.

This work was supported by Polish Ministry of Science and Higher Education within research grant PBZ-KBN-115/T08/04.

References

[1] R. M. Bozorth: Ferromagnetism (D. van Nostrood, New York 1956).

[2] D. C. Jiles: Introduction to Magnetism and Magnetic Maerials (Chapman&Hall, London 1998).

[3] A. Bienkowski, J. Kulikowski: J. Magn. Magn. Mater. Vol. 26 (1982), p. 292.

[4] T. Somdev: J. Appl. Phys. Vol. 55 (1984), p. 2464.

[4] R. Szewczyk, A. Bieńkowski, R. Kolano: Cryst. Res. Technol. Vol. 38 (2003), p. 320.

[5] H. Chiriac, E. Hristoforou, M. Neagu, I. Darie, C. Hison: Sensors and Actuators A Vol. 81, (2000), p. 150.

[6] H. Chirac, E. Hristoforou, M. Neagu, F. Barariu, I. Darie: Sensors and Actuators A Vol. 81 (2000), p. 147.

[7] J. Salach, A. Bieńkowski, R. Szewczyk: Journal of Automation, Mobile Robotics and Intelligent Systems, Vol 1. (2007), p. 66.

[8] A. Bieńkowski, R. Szewczyk: Sensors and Actuators A113 (2004), p. 270.

[9] R. O'Handley: Modern magnetic materials – principles and applications (John Wiley & sons, 2000).

[10] A. Bieńkowski, R. Szewczyk, Polish Patent P-345758, 2001.

Solid State Phenomena Vol. 154 (2009) pp 29-33
online at http://www.scientific.net

Measurements of strain in ceramic components using magnetostrictive delay line

Roman Szewczyk[1,a], Jacek Salach[1], Adam Bieńkowski[1],
Marek Kostecki[2,b], Andrzej Olszyna[2], Aleksandra Kolano-Burian[3,c]

[1] Institute of Metrology and Biomedical Engineering, Warsaw University of Technology, sw. Andrzeja Boboli 8, Warszawa 05-525, Poland

[2] Faculty of Materials Science and Engineering (InMat), Warsaw University of Technology, Wołoska 141, Warszawa 02-507, Poland

[3]Institute of Nonferrous Metals, Sowińskiego 5, Gliwice 44-100, Poland

[a]szewczyk@mchtr.pw.edu.pl, [b]mkostecki@o2.pl, [c]olak@imn.gliwice.pl

Keywords: magnetostrictive delay line, amorphous alloys, ceramics.

Abstract. Paper presents a novel application of magnetostrictive delay lines, which give a possibility of real time monitoring of strain in ceramic components. Magnetostrictive delay line was based on highly magnetostrictive Fe-Si-B amorphous alloy ribbon, mounted outside of ceramic component, what is a new solution for increasing sensor's sensitivity. Developed specially for this sensor, hybrid digital-analog signal processing unit covers the sample-and-hold integrated circuit. The achived sensitivity and repeatability of the sensor confirms, that such solution is suitable for ceramic machine tool monitoring.

Introduction

Ceramic components are used in different applications in machine industry. However, the lack of a real-time method for monitoring of the cracks in such ceramic elements remains a very significant hindrance, from the technical point of view. Therefore, development of a cost-effective and sensitive method of monitoring of ceramic parts may create new possibilities of application of smart material and machine structures for industry.

Magnetostrictive delay lines are sensitive for stresses from external forces [1, 2]. Principles of operation of such delay line are based on the fact, that the magnetostrictive impulse, generated by the rapid change of magnetization of driving coil, is moving in magnetostrictive material with the speed of sound [3]. In that case the magnetoelastic pulse may be detected as changes of voltage in detecting coil. The parameters of delay line (covering mechanical stresses) have significant influence on both strength, as well as time of signal reception in detecting coil [4].

Soft magnetic material used for the magnetostrictive delay line has strong influence on its functional parameters. For such lines materials with significant saturation magnetostriction λ_s are used, where the saturation magnetostriction λ_s can be determined as the value of magnetostriction in the sample, during its magnetization from demagnetized state to magnetic saturation [5]. Moreover, from the practical point of view, the magnetostrictive delay line has to be produced in form of a thin wire or ribbon [6]. For this reason, highly magnetostrictive intermetallic materials, such as Terfenol-D with saturation magnetostriction exceeding 1000 μ/m, are not suitable for magnetostrictive delay line. Soft amorphous alloys, in both ribbon and wire forms, seem to be the most useful in this case. Particularly, the iron-based alloys with saturation magnetostriction λ_s about 35 μm/m give new possibilities for magnetostrictive delay lines. These alloys in as-quenched state exhibit the high stress strength (about 1000 MPa [7]), significant rust resistance [8], and their production costs are much lower than the high permeability cobalt-based alloys.

Experimental setup

For the development of magnetoelastic delay line sensor, the ribbon made of $Fe_{78}B_{13}Si_9$ amorphous alloy was used. Amorphous alloy ribbon was produced by Institute of Non-ferrous Metals and was made in as-quenched state to reduce its fragility.

Experimental setup for testing the functional properties of developed sensor for monitoring of the ceramic components is presented in figure 1. Ceramic bar (1) is supported on its ends, whereas it's being loaded in the centre by the force F. Magnetostrictive delay line (2) is mounted outside of the bar, to create the sufficient conditions for magnetoelastic wave propagation. Moreover, in this solution the value of strain in magnetostrictive delay line is higher than the strain in ceramic rod. Magnetoelastic pulse is generated by the driving coil (3) and detected by the sensing coil (4). Both driving coil signal, as well as signal from the sensing coil are connected to electronic signal processing unit (5).

Force F applied to the ceramic rod is generated by the hydraulic press. The maximal value of force F exceeds 2000 N, and is measured by the strain-gauge transducer with 0.5% accuracy. As a result, ceramic rods with different sizes and made of different materials may be tested for wide range of stresses.

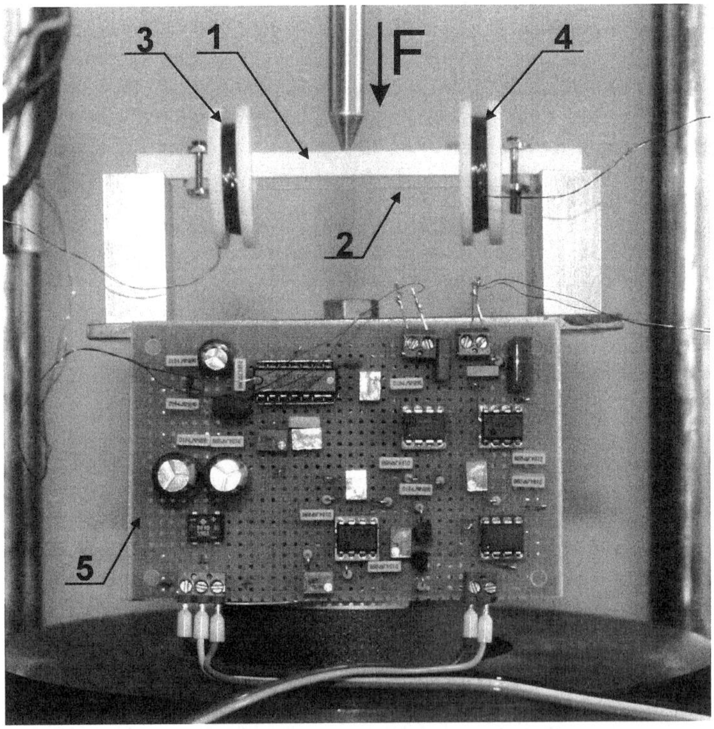

Fig. 1. Measuring setup for testing the magnetostrictive delay line stress sensors for ceramic components: F – loading force, 1 – ceramic component (bar), 2 – magnetostrictive delay line, 3 – driving coil, 4 – sensing coil, 5 – electronic circuit board

It should be indicated, that value of force F applied to the ceramic rod may be easily converted to its strain ε using the following equation:

$$\varepsilon = F\frac{l \cdot y_{max}}{4 \cdot E \cdot J_z} \quad (1)$$

where l is the length between supporting points of the rod, E is Young modulus of the material, J_z is moment of inertia of cross section, and y_{max} is the distance between neutral line and the farthest point of rod's cross section.

Electronic signal processing

A schematic block diagram of sensor's electronic signal processing unit is presented in figure 2. Signal processing unit was mounted on universal PCB to simplify the assembly.

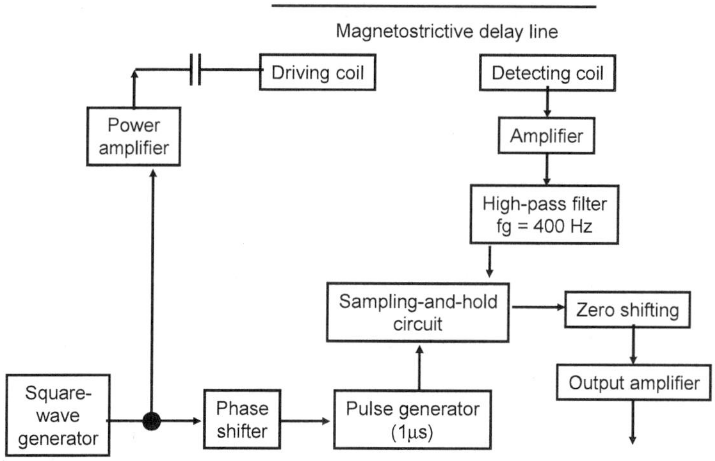

Fig. 2. Schematic block diagram of the electronic circuit of the magnetostrictive delay line stress sensor

Square wave generator generates signal of about 100 Hz frequency. On the positive edge of the square wave, the power amplifier generates magnetoelastic pulses (by the driving coil) in the magnetostrictive delay line. The signal from the driving coil is amplified, and the low frequency harmonics (lower than 400 Hz) are filtered out by the high-pass filter. Sampling-and-hold circuit lock value of signal from detecting coil in precisely given moments, optimal from the point of view of signal, changes under stresses. The time of sampling-and-hold circuit opening is shorter than 1 μs. As a result, the value of hold voltage is provided to zero shifting circuit and the low-pass output amplifier. It should be indicated, that frequency of square-wave signal shouldn't be to small, with regard for the drop of voltage in the sample-and-hold circuit. Whereas such drop is typically lower than 0.01 mV/ms, the frequency of square wave of about 100 Hz is sufficient from sensor operation point of view.

Results

In figure 3 the plots of high-pass filtered signal from the detecting coil, as well as the plot of difference between signal from stressed and unstressed magnetostrictive delay line is presented, for given value of force F applied to the ceramic bar. It can be observed, that under the stresses, the measured amplitude of magnetostrictive pulse is lower, as well as the pulse reaches the detecting coil later. Using the plot of difference between signal from stressed and unstressed magnetostrictive delay line one can determine the optimal point of sampling (marked by vertical broken line). Sampling in that point enables achievement of the maximal sensitivity of the sensor. Precise sampling in point determined that way may be achieved by changing the value of phase shifting (using the digital signal phase shifter as shown in figure 2).

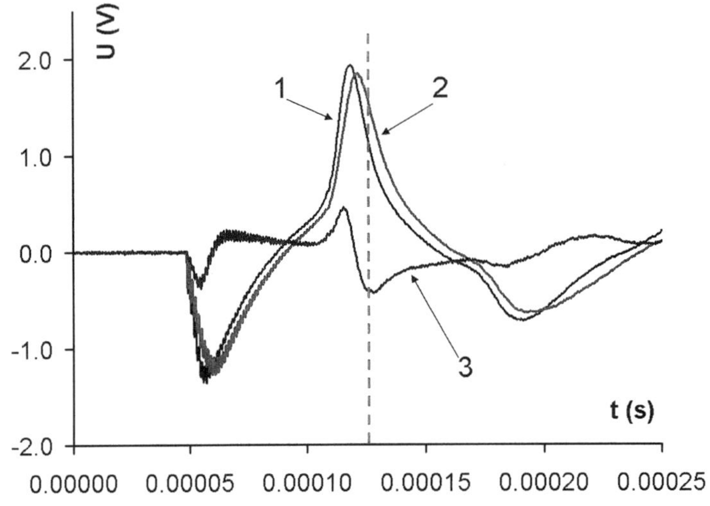

Fig. 3. Plots of signal on the detecting coil: unstressed delay line (1), under stresses generated by force 160 N (2), difference between signal from unstressed and stressed delay line (3)

In figure 4 the output characteristic U_{out} (F) of developed sensor is presented. When the value of sampling point is correctly determined, the sensor can achieve sensitivity higher than 31 mV/N, as well as standard deviation of repeatability of measurements equal 160 mV, what is sufficient from technical requirements point of view.

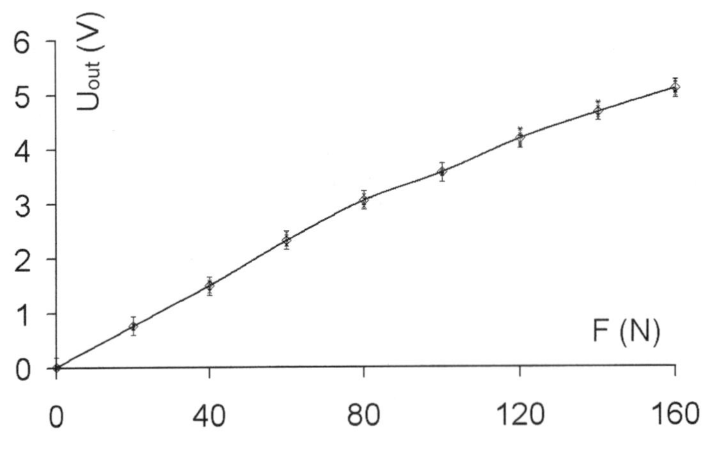

Fig. 4. Output characteristic of developed sensor utilizing magnetostrictive delay line for monitoring of stresses in ceramic components

Conclusion

Presented results confirm, that the developed method of real time monitoring of stresses in ceramic components exhibits high sensitivity, as well as repeatability suitable for technical applications. Force sensor based on novel solution, where magnetostrictive delay line is mounted outside of the ceramic bar, exhibits sensitivity higher than 31 mV/N. Moreover, the presented methodology of processing the signal from the magnetostrictive delay line enables development of cost-effective sensing devices, which may be integrated in larger structures, adequately to the demands.

This work was supported by Polish Ministry of Science and Higher Education within research grant PBZ-KBN-115/T08/04.

References

[1] E. Hristoforou, H. Chiriac, M. Neagu, L. Darie, T. A. Ovari, T.A.: IEEE Trans. Magn. Vol. 32, (1996) p. 4953

[2] E. Hristoforou, R. E. Reilly: IEEE Trans. Magn. Vol. 27 (1991) 5244

[3] H. Chiriac, E. Hristoforou, M. Neagu, M.Vazquez, C. Hison: IEEE Trans. Magn. Vol. 36 (2000) p. 3436

[4] E. Hristoforou, R. E. Reilly: IEEE Trans. Magn. Vol. 30 (1994) p. 2728

[5] E: Tremolet: Magnetostriction, CRC Press, London, 1992

[6] R. Hornreich, H. Rubinstein, R. Spain: IEEE Trans. Magn. Vol. 7 (1971) p. 29

[7] Information on http://www.metglas.com

[8] R. O'Handley: Modern magnetic materials – principles and applications, John Wiley & sons, 2000

Solid State Phenomena Vol. 154 (2009) pp 35-40
online at http://www.scientific.net

Magnetomechanical properties of Terfenol-D powder composites

Jerzy Kaleta, Daniel Lewandowski, Rafał Mech, Paweł Gąsior

Wroclaw University of Technology, Institute of Materials Science and Applied Mechanics, Smoluchowskiego 25, Wrocław 50-370, Poland

jerzy.kaleta@pwr.wroc.pl

Keywords: magnetostriction, experimental mechanics, smart materials, Terfenol-D composites

Abstract. In this work the methodology of determining mechanical and magnetomechanical properties (mainly magnetostriction) of original composite material on epoxy resin matrix and powdered Giant Magnetostrictive Material (Terfenol-D) has been presented. Researches were carried out in order to differentiate properties of the GMM composite with two different volume fraction of Terfenol-D powder (35% and 45%) in comparison with bulk Terfenol-D. Results show that composites magnetostriction highly depends on volume fraction of Terfenol-D powder, and it is few times lower than in monolithic material. Moreover a possible ways of application were indicated.

Introduction

Nowadays researches over new kind of materials mainly concern over so called SMART materials (SM). Generally speaking by SMs we mean such materials whose main utility characteristics expressed by means of the magnitude of one physical field (for instance mechanical field) strongly depend on the magnitude of another field, e.g. magnetic, electric or temperatures field. In this group an important role is played by Smart Magnetic Materials (SMM), some of which are the object of the present study and are described below. They include magnetorheological materials, materials with gigantic magnetostriction, materials of gigantic magnetoresitivity, magnetocaloric materials and others [1]. Giant Magnetostrictive Materials – GMM allow transformation of magnetic energy into mechanical energy (action of the actuator type) and mechanical energy into magnetic energy (action of the sensor type) with a very big efficiency coefficient, which enables multiple applications in many fields of technology.

Brittleness of Giant Magnetostrictive Materials is the source of various difficulties in their applications, which is why mainly cases of pressing loads are preferred. Increase in the work frequency induces eddy current [2], that is why there is frequency work limitations for this materials in the range of 10 to 20 kHz, but the useful bandwidth is below 1 kHz. This feature limits the material's usefulness in dynamic applications. Relatively high price of these materials is also of vital importance. For the past few years research projects have been implemented, whose aim is to develop new materials in the shape of various magnetostrictive composites, among others rods and laminates [3]. The matrix is usually epoxied resin reinforced with magnetostrictive material in the shape of powder, flakes or tiny rods usually made of Terfenol-D [4].

Material fabrication

In the study magnetostrictive composite (hereafter referred to as GMM composite) developed in the Institute of Materials Science and Applied Mechanics in Wroclaw University of Technology was used. The composite was fabricated through blending together epoxide resin and powdered GMM material – powder of Terfenol-D ($Tb_{(0.3)}$ $Dy_{(0.7)}$ $Fe_{(1.9)}$).

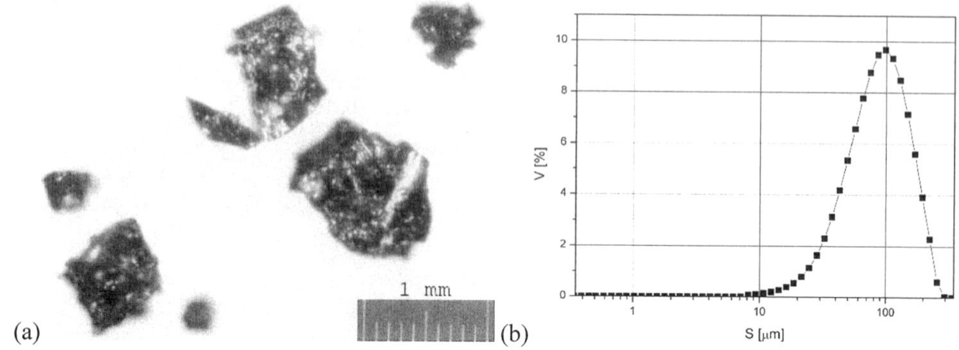

Figure 1. (a) Terfenol-D powder, used for fabrication of GMM composites and (b) - particle size distribution

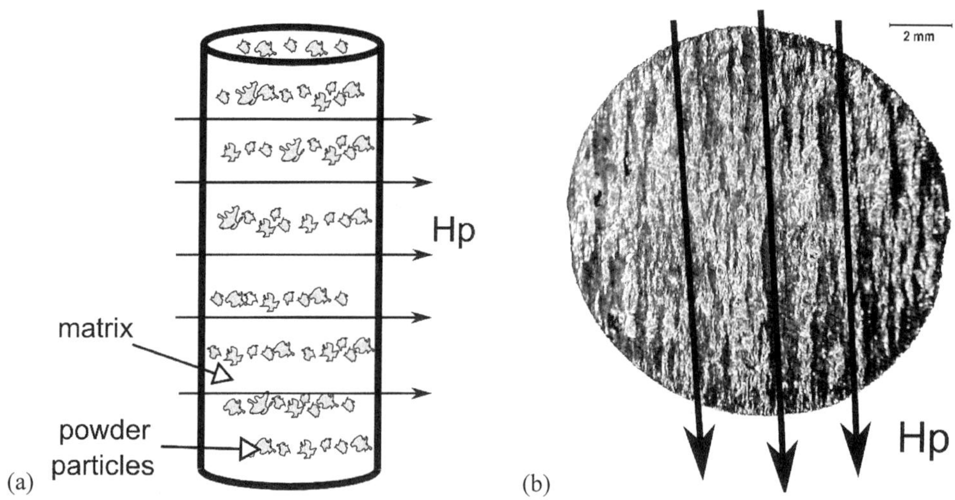

Figure 2. (a) - The scheme of polarization composites during last stage of fabrication process, direction of polarization field H_p prependicular to main sample axis. (b) - Transverse polished section of the created composite.

In the first stage of production process hardener and powdered Terfenol-D (producer: Gansu Tianxing Rare Earth Functional Materials Co., Ltd.) were added to the epoxide resin. Figure 1a presents pictures of particles powder used materials. It is seen there particles have different size and their shape is irregular, while their edges are sharp. Volume fraction distribution of used powder was shown in Figure 1b. Grain size was measured from 0 – 300 μm.

The mixture of resin and powder was exposed to the process of deaeration. The mixture was then poured into previously prepared containers of cylindrical shape and they were initially polarized, inserting samples between permanent magnets, and then once again it was put into a hermetically closed chamber in order to suck the air that accumulated in the process of mixing from it. Specimens were placed between two powerful magnets to effect their final polarization, perpendicular to the main axis of the specimen, and left until the resin had set completely. Scheme of polarization and direction of polarization field H_p were shown in Figure 2a.

Specimens of different volume content of Terfenol-D powder were prepared (35% and 45%). Figure 2b shows photo of the transverse polished surface of the specimen. It is possible to see on it chains made by particles of the powder. The strong polarization field was used to get anisotropy of the material and such a turn of the particles that their internal domain structure assumed its position along the newly created chains. Such an arrangement should ensure that, the magnetic field acting parallel to the main specimen axis will cause their turn and in consequence the occurrence of the stronger magnetostriction, than the one occurring in the case of particles placed at random [3]. The insertion of different volume content of Terfenol-D in the composite, served to check the influence of it on the magnetomechanical properties of the material.

Experimental setup

Original stands for determining magnetomechanical properties of GMM composites have been set up. The main part of the stand was an electromagnetic circuit consisting of a coil and metal casing. The diameter of the coil opening in which the specimen was placed, was chosen in such a way that it increased the acting magnetic field to the maximum, and at the same time it was possible to place a measuring apparatus in the form of the fiber optic strain sensor and temperature and magnetic field sensors in the opening. The scheme of the experimental stand and its measuring part has been presented in Figure 3a. The range of the magnetic field intensity used during the research was within the range between of 0 to 180 kA/m.

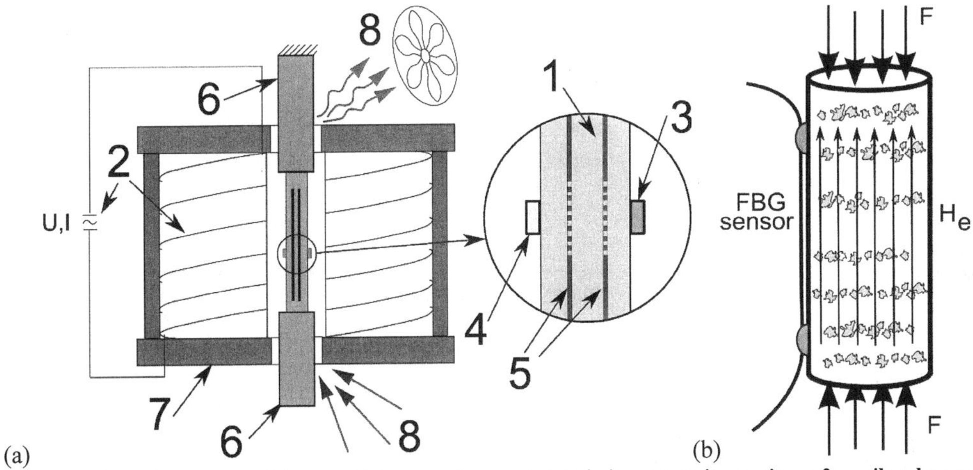

Figure 3. Measuring scheme of the stand (a), where 1 – magnetostrictive composite specimen, 2 – coil and power supply , 3 – magnetic field sensor, 4- temperature sensor, 5 – strain sensors (Bragg grating), 6 – hydraulic pulsator MTS , 7 – steel casing, 8 – air cooling system. (b) – F loading specimens direction, magnetic field vector applied during experiment H_e and idea of mounting FBG sensor on outer surface

For the sake of experiment conditions (sample shape, size and range of magnetic field strength) for strain measurement was used Fiber Bragg Grating sensor (FBG) [5]. In comparison to ordinary methods, like strain gauges, FBG sensors are insensitive on magnetic field and easy to compensate influence of temperature. Precision, sensitivity and range of measurement was useful for register small changes of magnetostriction [6]. Glass fibers with Bragg grating were mounted in two points glued for the surface as shown in Figure 3b. Two FBG sensors were used for one measurement, first one for measurement strain, second for temperature compensation.

Specimens loading methods are shown in Figure 3b. During experiment external magnetic field H_e was perpendicular to direction magnetic field during specimens preparation and parallel to direction of loading and strain measurement. This arrangement was undertake to obtain highest magnetostriction effect, which was dependent on angle of magnetic domain rotation, therefore their position in zero magnetic field[3].

Mechanical and magnetomechanical examinations

Composite specimens have been examined in relation to their mechanical and magnetomechanical properties. In the first place Young modul was determined, crucial in each constitutive material model. E modul was determined for each of the fabricated composites, as well as a specimen made of pure epoxide resin and a specimen of monolithic Terfenol-D. The examinations consisted in exposing of composite specimens to static pressing without stimulation with external magnetic field. The obtained results are presented in Figure 5.

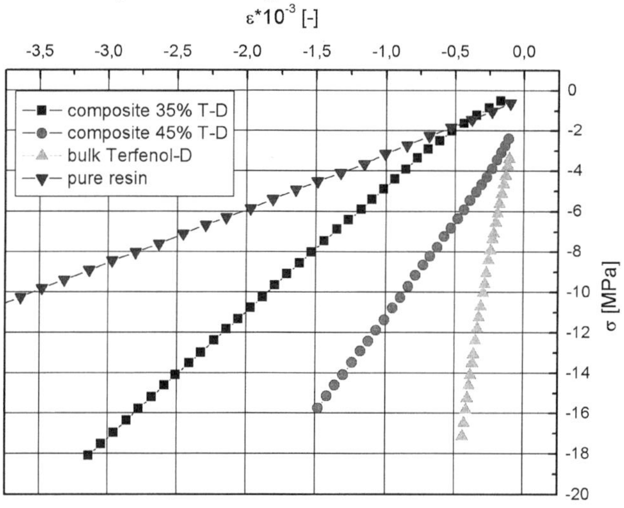

Figure 5. Static tests of loading (compression) for composite specimens and the specimen of pure resin (matrix without filling) and solid Terfenol-D rod.

As shown above Young modul depends largely on the volume content of Terfenol-D, which seems obvious, considering the differences in hardness and durability of bulk Terfenol-D and resin. The subsequent examinations of the composites produced regarded their magnetostrictive properties. For this purpose composite Terfenol-D was examined, and – in order to compare the results obtained – monolithic specimen of Terfenol-D. Measurement was made for a constant prestress load $\sigma_o = 7$[MPa]. Changes in the magnetic field were measures by the magnetic field sensor placed near the specimen in a space between the coil and the specimen. The Bragg grating sensor described above measured elongation of the specimen.

Figure 6 presents the results of the tests for composites with the filling of 35% and 45% and for the monolithic specimen of Terfenol-D. It is seen clearly that the obtained results for composites are decidedly smaller in comparison with monolithic material, what is similar with the results achieved in works [7, 8]. This can be caused by the interaction of Terfenol-D powder and composite matrix. Although the value of the magnetostriction for GMM composite is several dozen per cent lower, it is still about 30 times higher than for nickel, which is in this case our reference point. Development of the composite, has in turn led to a significant reduction in its brittleness and price in comparison with solid Terfenol-D.

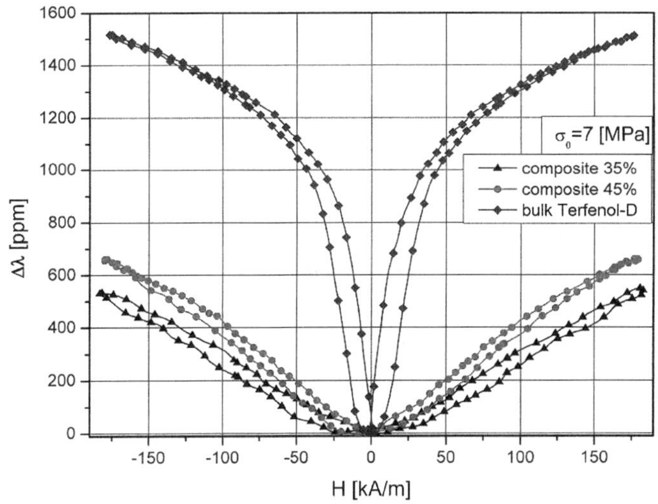

Figure 6. Dependence of magnetostriction on the change in magnetic field intensity H, at initial stress 7MPa, for composites 35% and 45% and monolithic Terfenol-D

Generally the initial findings are so promising that they should result in a wider program of research and further examples of these materials applications especially in the scope of higher frequencies, where solid Terfenol-D will be replaced by composites (e.g. active dampers, actuators etc.) Furthermore initial studies make it possible to develop a constitutive model for GMM composite.

Summary

New technology of fabricating an original composite on the matrix of epoxide resin and powdered material of gigantic magnetostriction (Terfenol-D) has been worked out. Magnetic polarization that was used made it possible to obtain programmed anisotropic characteristics of the material. The composite that was created as a result, meets the requirements laid down for magnetic materials from the Smart group, which means that the change in mechanical properties may be effected, in the wide range, through external stimulation with the magnetic field. Methodology of examining magnetomechanical properties of GMM composite under conditions of controlled changes in magnetic and mechanical parameters at controlled work temperature has been worked out. Measurement of the deformation of the composite rod with Bragg gratings (FBG) is a quite new concept. It has been proved that GMM composite is cheaper, has weaker eddy currents and is less fragile in comparison with Terfenol-D, which gives it wider range of applications. It is true that magnetostriction is smaller than for solid Terfenol-D, but it is still tens of times bigger than in the case of nickel.

References

[1] „*Encyklopedia of Smart Materials*", John Wiley&Sons, 2002, editor: M.Schwartz.

[2] D. Kendall, A. R. Piercy, „*Terfenol-based materials with reduced eddy-current loss for high-frequency applications*'' Ferroelectrics, Volume 187, pages 153-161, year 1996.

[3]G.P. McKnight „*[112] Oriented Terfenol-D Composites*" PhD Thesis, Universit f California, 2002.

[4] C. Yin Lo, S. Wing Or, H. Lai Wa Chan. „*Large Magnetostriction in Epoxy-Bonded Terfenol-D Continous-Fiber Composite With [112] Crystallographic Orientation.*" IEEE TRANSACTON ON MAGNETICS, 42,2006

[5] H. L. Liu, H. Y. Tam, Ching Yin Lo, Siu Wing Or „*A novel high frequency magnetostrictive composite-fiber Bragg grating sensor*" Optical Fiber Communication and the National Fiber Optic Engineers Conference, 2007. OFC/NFOEC 2007.

[6] F. Yu , S. Yin: *Fiber Optic Sensors*; Marcel Dekker Inc., 2002.

[7] G. Altin, K. K. Ho, C. P. Henry, G. P. Carman „*Static properties of crystallographically aligned Terfenol-D/polymer composites.*" Journal of Applied Physics, Volume 101, year 2007.

[8] N. Nersessian, S. Wing Or, G.P. Carman „*Magneto-thermo-mechanical characterization of 1-3 type polymer-bonded Terfenol-D composites*'' Journal of Magnetism and Magnetic Materials, 263, 2003, pages 101-112.

Solid State Phenomena Vol. 154 (2009) pp 41-46
online at http://www.scientific.net

A Simulation Study of Magnetostrictive Material Terfenol-D in Automotive CNG Fuel Injection Actuation

H.A. Chowdhury[1, a], S.A. Mazlan[1,b] and A.G. Olabi[1,c]

[1] School of Mechanical & Manufacturing Engineering, Dublin City University, Dublin 9, Ireland.

[a] chowdhh2@mail.dcu.ie, [b] saiful.mazlan2@mail.dcu.ie, [c] abdul.olabi@dcu.ie

Keywords: Magnetostrictive, ANSYS, FLUENT, CNG Fuel Injector

Abstract. Magnetostriction is the deformation that spontaneously occurs in ferromagnetic materials when an external magnetic field is applied. In applications broadly defined for actuation, magnetostrictive material *Terfenol-D* ($Tb_{0.3}Dy_{0.7}Fe_{1.9}$) possesses intrinsic rapid response times while providing small and accurate displacements and high-energy efficiency. These are some of the essential parameters required for fast control of fuel injector valves for decreased engine emissions and lower fuel consumption compared with the traditional solenoid fuel injection system. A prototype CNG fuel injector assembly was designed which included magnetostrictive material Terfenol-D as the actuator material. A 2D cross-sectional geometry of the injector assembly, which incorporated both linear and non-linear magnetic properties of the corresponding materials, was modeled in *ANSYS* for 2D axisymmetric magnetic simulation. Subsequently, a 3D replica of the CNG flow conduit was modeled in *GAMBIT* with the resultant injector lift. The meshed conduit was then simulated in *FLUENT* using the 3D time independent segregated solver with the *Standard* $k - \varepsilon$, the *Realizable* $k - \varepsilon$ and *RSM* turbulence models to predict the mass flow rate of CNG to be injected. Eventually, the simulated flow rate was verified against mathematically derived static flow rate required for a standard automotive fuel injector considering standard horsepower, BSFC and injector duty cycle.

Introduction

In many of the emerging engine technologies inspired by emission reduction legislations, it is becoming evident that the integration and adoption of multi-functional smart materials are essential alternatives for the development of fuel injection applications. The integration of actuating capability at the micron scale affords the realization of extremely controlled and effective fuel injector actuator system that would be able to respond to an environmental stimulus such as change in magnetic field. *Magnetostriction* (MS) is such a phenomenon in which the shape of certain type of materials changes under the influence of an external magnetic field [1]. This change in shape is the result of the rotation of small magnetic domains which causes internal strains in the material structure. The magnetostrictive effect was first described in the 19th century (1842) by an English physicist James Joule.

Traditional solenoid fuel injection system suffered with lack of ability to operate under high speed applications mainly due to factors such as inertia; resulting in wear in valve seat, needle noise and unwanted vibrations. Thus, the exquisite magnetostrictive actuating capabilities at micron scale and the incompetence of traditional solenoid fuel injection system motivated this research to discover the possibility of implementing a magnetostrictive material, in this case *Terfenol-D*, in automotive fuel injection actuation.

Prototype Fuel Injector

A prototype fuel injector assembly was designed using *Pro/Engineer Wildfire 3.0*, which is a 3D CAD parametric feature solid modeling software created by Parametric Technology Corporation (PTC), as shown in Fig. 1. A *Terfenol-D* shaft (30mm x 50mm) was used as the core actuator material. Besides, the injector housing was composed of *1020 Steel*, which is a low carbon steel having soft magnetic material properties.

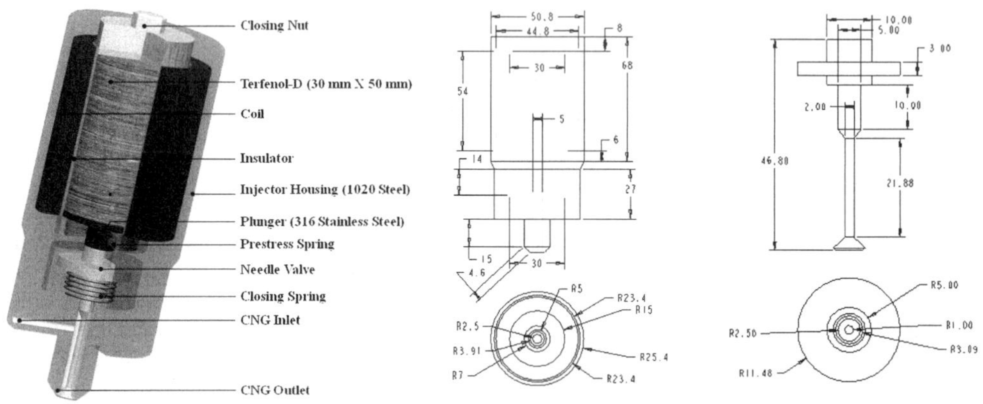

Injector Assembly | Injector Housing (dimensions in mm) | Needle Valve (dimensions in mm)

Fig. 1. Prototype Fuel Injector

In order to achieve an injector lift of 50 Microns, a magnetic field strength or Flux Density of about 0.55 Tesla had to be generated by the coil in the Terfenol-D region. In this paper, an ANSYS 2-D static magnetic nodal simulation would investigate whether this magnetic field strength could be achieved for a coil thickness of 5 mm comprising 22 AWG copper coil wire with 672 turns in 8 layers. Further discussion regarding these obtained parameters could be found from Chowdhury, et al. [2]. Furthermore, a *computational fluid dynamics* (CFD) simulation will be implemented to predict the mass flow rate of the prototype fuel injector.

ANSYS 2D Static Electromagnetic Nodal Analysis

Element Type. The two-dimensional injector model was constituted by *PLANE13*, which was a two dimensional, coupled solid element type having up to four degrees of freedom per node. These degrees of freedom could be magnetic vector potential (AZ), displacements, temperature or time-integrated electric potential. Furthermore, PLANE13 had a 2-D magnetic, thermal, electrical, piezoelectric and structural field capability with limited coupling between the fields. The element behavior was set as axisymmetric.

Magnetic Formulation. The ANSYS Electromagnetic simulation incorporating PLANE13 element type was solved with nodal based *magnetic vector potential* (MVP) formulation method, which has three more degrees of freedom (DOF) per node than the scalar method. These are magnetic vector DOFs in the X, Y and Z directions, namely AX, AY and AZ. Voltage-fed or circuit coupled analysis may add up to three additional degrees of freedom to the magnetic vector DOFs. These are namely *current* (CURR), *electromotive force drop* (EMF), and *electric potential* (VOLT).

ANSYS Electromagnetic Results. After modeling the 2D injector geometry, and implementing the mesh size controls for the corresponding material areas, a quadratic mesh was generated throughout the injector cross-sectional area. The outside lines of the geometry were selected as the magnetic vector potential boundary and the current density was applied as magnetic excitation. Fig. 2(a) shows the flux density distribution throughout the entire injector actuator by producing a half (½) expanded plot representation based on the defined axisymmetry. Moreover, Fig. 2(b) represents electromagnetic force calculations on element components, where the command *FMAGSUM* invoked an ANSYS macro that summarized the Maxwell forces.

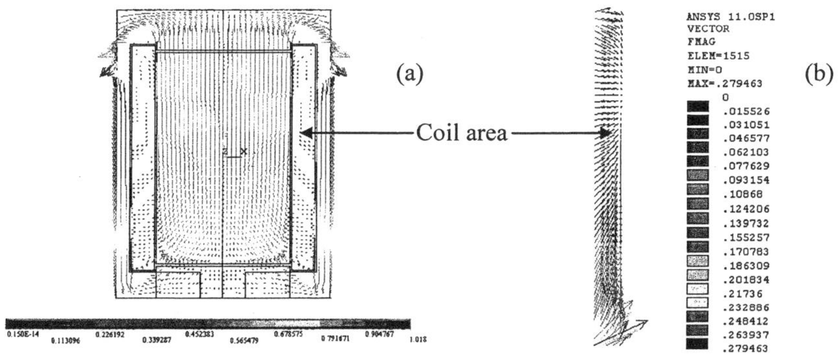

Fig. 2. (a) Flux density vector sum vector plot for 2D ½ expansion; (b) EMF vector plot.

Furthermore, as seen from the Fig. 3, the flux density distributions stayed more or less uniform along the different contour regions, implying no major fluctuations of flux density along the Terfenol-D member, which would result in maximum possible strain output.

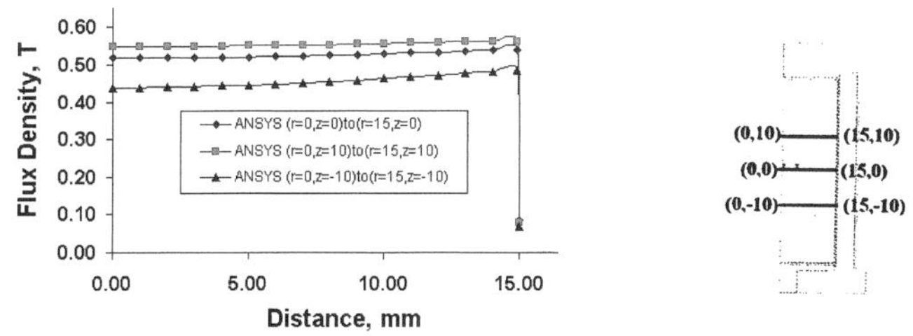

Fig. 3. Flux density distribution along the indicated contour lines.

Computational Fluid Dynamics Simulation

Modeling of the CNG Conduit. In the present study, *Standard k – ε*, *Realizable k – ε* and *RSM* turbulence models were utilized for the modeling of the CNG flow conduit to predict the mass flow rate of CNG for the injector lift of 50 Microns.

Geometry Creation. The 3D geometry of the CNG flow conduit was modeled in the *Pro-Engineer Wildfire 3.0* which was then imported in *Gambit version 2.4.6* as *IGES* format. The Gambit allows geometry to be constructed or imported from alternative packages as different formats, such as ACIS, IGES, STEP, Parasolid, Catia and CAD with cleanup and modification capabilities.

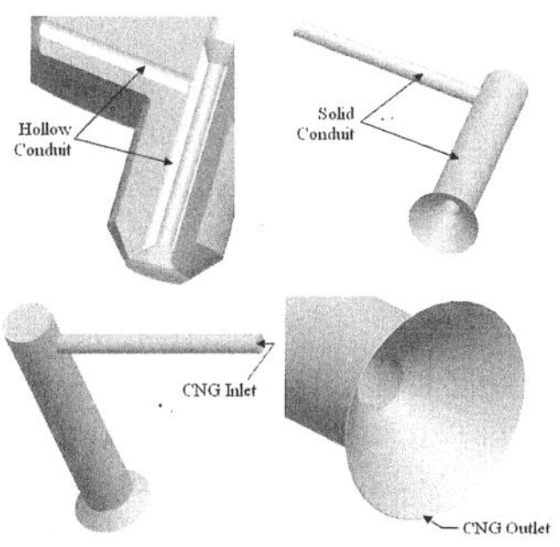

Fig. 4. Pro-Engineer model of the CNG flow conduit.

The Pro-E 3D solid model of the CNG flow conduit was constructed from the inner hollow channel of the original injector assembly having the same dimensions, as shown in Fig. 4. Eventually the model was cleaned by using *Virtual Cleanup* tool in GAMBIT. Furthermore, specifying the *no-stand alone edges* enabled GAMBIT not to read in any edges that do not belong to faces or volumes. In this research, *FLUENT 5/6 solver* was selected in GAMBIT version 2.4.6 for CFD analysis.

Specifying Zone Types. Zone-type specifications generally define the physical and operational characteristics of the model at its boundaries and within specific regions of its domain. There are two classes of zone-type specifications, namely *Boundary types* and *Continuum types*. Where *Boundary-type specifications* defined the physical and operational characteristics of the model at those topological entities that represent model boundaries, *Continuum-type specifications* defined the physical characteristics of the model within specified regions of its domain. In this research, the CNG inlet and CNG outlet face regions were consecutively defined as *pressure inlet* and *pressure outlet* boundary types. Furthermore the rest of the faces were defined as *wall boundary type* and the volume was defined as the *continuum type* which was *FLUID* in this particular case.

Meshing. The meshing of the model geometry is an important and challenging task because the quality of the mesh affects the accuracy of the solution, the CPU time and memory requirements [3]. In this simulation, the *Mesh Volumes* command allowed to create the mesh for the entire volume of the model. When meshing the volume, GAMBIT created mesh nodes throughout the volume according to the specified meshing parameters. In order to specify the meshing scheme, two parameters were defined. These are *Elements* and *Type*. The *Elements* parameter defined the shape of the elements that were used to mesh the volume. In this case, a *Tet/Hybrid* element type was chosen which specified that the mesh was composed primarily of tetrahedral mesh elements but may include hexahedral, pyramidal and wedge elements where appropriate. Subsequently, the *TGrid* type was associated with the *Tet/Hybrid* element type. The *TGrid* meshing scheme enabled GAMBIT to create an array of pyramidal mesh elements in proximity of the faces and furthermore creating tetrahedral elements throughout the rest of the volume.

Mesh Checking. It is important to check the quality of the resulting mesh, because properties such as skewness can greatly affect the accuracy and robustness of the CFD solution. GAMBIT provides several quality measures which can assess the quality of the mesh generated. Table 1 describes the resulting mesh characteristics based on corresponding quality types.

Table 1: GAMBIT mesh investigation.

Quality type	Highest active elements (%)	Range	
		Lower limit	Upper limit
EquiAngle Skew	95.24%	0	0.75
EquiSize Skew	95.26%	0	0.75

An important factor regarding mesh checking was to ensure that the equi-angle skewness was not to exceed the value 0.75 [4]. As seen from Table 1, a high percentage of the 3D elements were within this range. For an ideal mesh, in the statistical distribution of mesh elements represented by a histogram, the bars on the left of the histogram would be large and those on the right would be small. In this case, each element would possess a value of skewness between 0 and 1, where 0 would represent an ideal element.

FLUENT Simulation Results. The purpose of the CFD simulation was to determine the mass flow rate of CNG at the pressure outlet of the fuel injector. Subsequent to reading the **.mesh** file from GAMBIT, FLUENT created the **.case** and **.data** file decipherable for CFD analysis. The pressure at the inlet and the outlet of this flow conduit could then be varied depending on whether the injector to be utilized in case of either port injection or direct injection applications. In this research, in order to compare the resultant simulated mass flow rate with a standard intake port injected natural gas engine having identical fuel injection operating conditions, the following assumptions were made, as shown in Table 2. As mentioned earlier, the FLUENT simulations were carried out for three different turbulent models, namely *the Standard k–ε, the Realizable k–ε* and *RSM* models.

Table 2: Assumptions for CFD analysis.

Turbulence Model	Standard k–ε Model		
	Realizable k–ε Model		
	Reynolds Stress Model (RSM)		
Solver	Segregated	***Formulation***	Implicit
		Time	Steady
		Space	3D
		Velocity Formulation	Absolute
Boundary Conditions	***Pressure Inlet (psig)***	85	
	Pressure outlet (psig)	0	
	Wall	Stationary, No slip	
	Fluid	Methane (CH_4)	

Subsequently, Fig. 5 shows the static pressure distribution at the outlet region of the fuel injector for all three turbulent models.

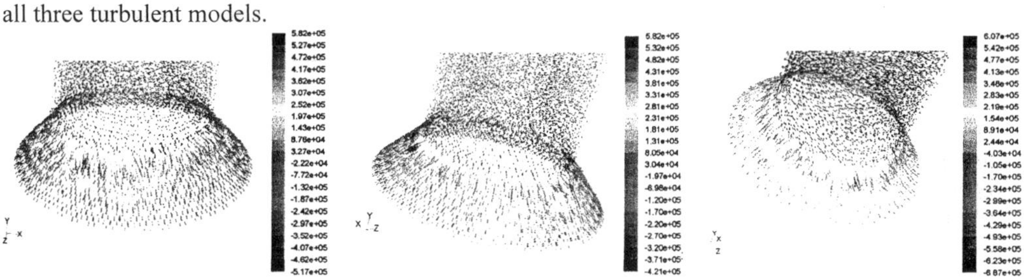

FLUENT 6.2 (3d, segregated, standard k-ε) *FLUENT 6.2 (3d, segregated, realizable k-ε)* *FLUENT 6.2 (3d, segregated, RSM)*

Fig. 5. Velocity vector colored by static pressure (Pa).

Sample Calculation of Injector Flow Rate. The *brake specific fuel consumption* (BSFC) of an automotive engine is a measure of engine efficiency expressed as the rate of fuel consumption divided by the power produced, which could be represented as,

$$B.S.F.C = \frac{Fuel\_Flow\_Rate}{Horsepower} \tag{1}$$

Considering the *duty cycle* of the fuel injector, the required injector static flow rate at 43.5 psig was [5]:

$$Injector\_Static\_Flow\_Rate_{43.5psig} = \frac{Horsepower \times B.S.F.C}{Injector\_Number \times Duty\_Cycle} \tag{2}$$

Again, in order to find the new fuel flow rate after the change of fuel pressure [5]:

$$New\_Flow\_Rate = \sqrt{\frac{New\_Pressure}{Old\_Pressure}} \times Old\_Fuel\_Rate \tag{3}$$

The assumptions for these calculations are shown in Table 3.

Table 3: Assumptions for sample fuel flow calculation

Entity	Value	Remark
B.S.F.C	0.50	For naturally aspirated engine (0.45-0.50)
Duty Cycle	0.80	Accepted industry standard for safe operation
Crankshaft Horsepower (BHP)	250	Assumption
Number of Injectors	6	Assumption
Old Pressure (psig)	43.5	Functional pressure for Eq. 2
New Pressure (psig)	85	Inlet pressure of fuel injector in CFD analysis

Eventually, the percentage differences between the simulated and numerically derived fuel injector mass flow rates are depicted in Table 4:

Table 4: Percentage difference between simulated and numerically derived mass flow rate for P_{inlet}: 85 psig and P_{outlet}: 0 psig

Turbulent model	Simulated mass flow rate (kg/sec)	Numerically derived mass flow rate (kg/sec)	Percentage difference (%)
Standard k-ε	0.00527488	0.0045	15.85%
Realizable k-ε	0.005214056		14.70%
Reynolds Stress Model (RSM)	0.005530861		20.55%

Conclusions

According to ANSYS Electromagnetic results, for a coil thickness of 5 mm with 22 AWG wire having 672 turns in 8 layers, possessed good prospect in actuating the desired amount of injector lift of 50 microns as the required magnetic field strength was achieved to cause the necessary amount of magnetostrictive strain. Furthermore, the purpose of the CFD analysis of the prototype fuel injector was to predict the mass flow rate of CNG at a specific operating condition. In this research, the analysis was performed to simulate the injection environment as if the injector was operated in case of an inlet port fuel injection system, where a standard port operated fuel injector inlet pressure was considered. Although the prototype injector model successfully operated under this simulated environment, the amount of mass flow rate at the pressure outlet was found to be slightly higher compared to the mathematically derived one. However, the mathematical model disregarded factors such as the cross-sectional area of injector outlet, turbulent flow and wall boundary conditions.

References

[1] L. Daniel, O. Hubert, and R. Billardon, Computational and Applied Mathematics, Vol. 23, Issue. 2-3 (2004), pp. 285-308.

[2] H. A. Chowdhury, S. A. Mazlan and A. G. Olabi, Implementation of Magnetostrictive Material Terfenol-D in CNG Fuel Injection Actuation, Advanced Materials Research, Vols. 47-50 (2008), pp. 630-633.

[3] P. L. George, Automatic Mesh Generation and Finite Element Computation, Handbook of Numerical Analysis, Vol. 5, Elsevier (1996), pp. 69-190.

[4] M. C. Bastarrica and H. N. Kaher, Designing a Product Family of Meshing Tools, Advances in Engineering Software (2005), pp. 1-10.

[5] Technical notes, RC Fuel Injection, Available from: www.rceng.com.

Solid State Phenomena Vol. 154 (2009) pp 47-52
online at http://www.scientific.net

SMA Thin Strip for Rotary-Driving Element

H. Tobushi[1,a], E. A. Pieczyska[2,b], W. K. Nowacki[2,c] and Y. Sugimoto[1,d]

[1] Department of Mechanical Engineering, Aichi Institute of Technology,.

1247, Yachigusa, Yakusa-cho, Toyota, 470-0392, Japan

[2] Institute of Fundamental Technological Research, Polish Academy of Sciences,

21 Swietokrzyska, Warsaw, 00-049, Poland

[a] tobushi@aitech.ac.jp, [b] epiecz@ippt.gov.pl, [c] wnowacki@ippt.gov.pl, [d] p07711pp@aitech.ac.jp

Keywords: Shape memory alloy, Thin strip, Torsion, Rotary driving element, Strain energy

Abstract. If a shape-memory alloy (SMA) thin strip is applied as an element subjected to torsion, a rotary driving element with a simple mechanism can be developed. The torsion tests were carried out for the SMA thin strip. Torque and recovery torque, both increase in proportion to the angle of twist and temperature. The recoverable strain energy increases in proportion to temperature. The dissipated work decreases slightly with an increase in temperature. A means of opening and closing a door with an element driven by an SMA thin strip is demonstrated.

Introduction

In recent years, the development and applications of smart materials have attracted worldwide attention. Shape-memory alloys (SMAs) have played a leading part in research of smart materials[1, 2]. The main properties of SMAs are the shape memory effect (SME) and superelasticity (SE). Thanks to these SMAs can be used in the driving elements of actuators, robots and heat engines. The SME and SE appear as a result of the martensitic transformation (MT).

The deformation properties of the SME and SE strongly depend on temperature and stress. Because of the adaptable thermal response of SMA elements, thin wires and thin tapes are widely used in practical applications[3]. In order to improve the flight performance of the rotor aircraft, twist of the blades was investigated recently by using the torsional deformation of a TiNi SMA tube [4].

In practical applications making use of SMA thin strips, torsional deformation can be obtained simply by holding both ends in the plates without any gripping device. If the characteristics of SE are exploited, a high performance of energy storage can be achieved similar to that of a torsion bar. In this way, simple and small actuators can be developed. In the case of the SMA thin strip twisted by fixing both ends, not only torsional but also tensile stress can be induced along both edges of the strip.

In the present paper, the torsional deformation properties of a TiNi SMA thin strip are investigated in a torsion test with loading and unloading at various temperatures. A means of opening and closing a door driven by an SMA thin strip is demonstrated.

Experimental Method

Materials and specimen. The materials used in the experiments were Ti-50.18 at% Ni SMA thin strips (NT-M2) with a thickness of t = 0.25 mm and a width of w = 5 mm produced by Furukawa Electric Co., Ltd. The specimen was a uniform flat tape of length L = 60 mm. The gauge length of the specimen was l = 40 mm and the length held in the gripping device was 10 mm at each site.

The as-received material was first hot-rolled from a wire with a diameter of d = 3.5 mm heated at 1073 K into a strip with a thickness t = 1.5 mm, followed by cold-stripping in several steps into the strip with a thickness of t = 0.25 mm. The hardening mechanism was supplied not by precipitation hardening but by work-hardening. A fat plane was shape-memorized by fixing in a flat plane for 1 min at 803 K in the furnace.

The as-received material was heat-treated again in an electric furnace to memorize the flat plane. In this heat-treatment, the as-received material was first fixed in a flat plane, heated up to 673 K for 2 hours and kept at 673 K for 1 hour, followed by cooling in the furnace. The same tests were carried out for both the as-received material and the heat-treated material used in the originally undeformed state.

Experimental procedure. The torsion and the recovery torque tests were carried out as follows.

Torsion test. The specimen was fixed onto the grips of the twisting shaft. The torque gauge was fixed on the twisting shaft and the specimen was twisted at the prescribed angle of twist per unit length θ (total angle of twist ϕ). The angle of twist and torque were measured simultaneously. In the torsion test, the maximum angle of twist per unit length was applied and thereafter torque was completely removed during unloading. The test temperatures T were room temperature (RT), 333, 343 and 353 K. The test was carried out in air at RT and in water at 333, 343 and 353 K. The temperature rise of the specimen during twisting in air was smaller than 4K. Therefore, the influence of the temperature rise is negligible. The maximum angle of twist per unit length θ_{max} was 78.5 rad/m (maximum total angle of twist $\phi_{max} = \pi$).

Recovery torque test. The specimen was first twisted at room temperature in air up to the prescribed angle of twist per unit length θ (total angle of twist ϕ). Torque was measured at this angle and the angle of twist was held constant. Maintaining this constant angle of twist, the specimen was soaked in water at the prescribed temperature, and torque was measured in the heated state. The recovery torque used under thermal cycling was determined by the difference of torque between the experimental and room temperatures. The temperature at which the specimen was twisted was room temperature, and the experimental temperatures T were 333, 343 and 353 K.

Experimental Results and Discussion

Deformed state in torsion. In the present study, the strip was twisted from one end while both ends were held constant at the axial position. Photographs of the deformed SMA thin strip are shown in Fig. 1 for each angle of twist. The left side in each case shows the fixed end and the right side shows the twisting end.

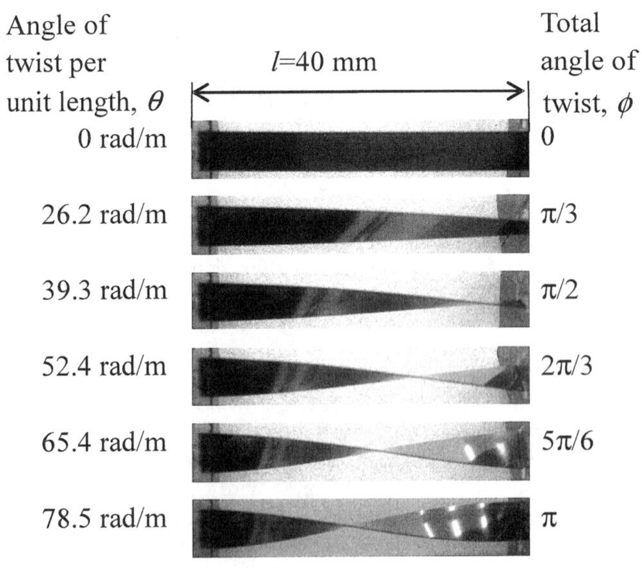

Fig. 1 Photograph of twisted SMA thin strip at each angle of twist.

As it can be seen, the upper and lower surfaces of the SMA thin strip are reversed during twisting and a crossover point appears in the twisting end at the angle of twist per unit length θ=39.3 rad/m (total angle of twist $\phi = \pi/2$). The crossover point propagates from the twisting end and reaches the central part of the specimen at the angle of twist per unit length with θ= 78.5 rad/m. Comparing the deformed states at θ= 0 and 78.5 rad/m (ϕ= 0 and π), both edges of the thin strip are found to be elongated by twisting since both ends are axially fixed. Therefore, tensile stress is induced along both edges and the stress state becomes much more complex than torsion in a cylindrical bar.

Relationship between torque and angle of twist. The relationship between torque M and angle of twist θ for the heat-treated material as obtained from the torsion test is shown in Fig. 2.

As it can be seen (Fig. 2a), M increases with an increase in θ. At room temperature, a large residual angle of twist per unit length appears after unloading, proving the SME. At T = 333 K, since T is $A_s < T < A_f$ and therefore there is a partial effect of superelasticity in which the reverse transformation is not completed. Residual angle of twist appears after unloading. At temperatures above 343 K, the angle of twist recovers during unloading and no residual angle of twist was noticed.

As can be seen from the relationship between torque and temperature in the loading process, shown in Fig. 2(b), torque M increases in proportion to temperature rise ΔT for the same angle of twist per unit length θ at temperature range above A_s. The slope increases in proportion to θ. This relationship is expressed as follows

$$M = \alpha_1 \theta (T - M_s), \tag{1}$$

where M_s denotes the MT starting temperature without stress and M_s = 305 K. This relationship appears owing to the fact that the MT stress increases proportionally to temperature (T-M_s).

In the case of torsion in a bar of round cross-section, with diameter d, the angle of twist ϕ is given by the equation, $\phi = Ml/GI_p$ where G and $I_p = \pi d^4/32$ are the modulus of rigidity and the polar area moment of inertia, respectively. For a bar of rectangular cross-section (width w and thickness t), the angle of twist per unit length θ is expressed as follows[5]

$$\theta = \frac{\phi}{l} = \frac{M}{k_1 w t^3 G}, \tag{2}$$

where the coefficient k_1 depends on the w/t ratio. Considering the fact that the M-θ curves during loading shown in Fig. 2(a) are close to straight lines, Eq. (1) becomes

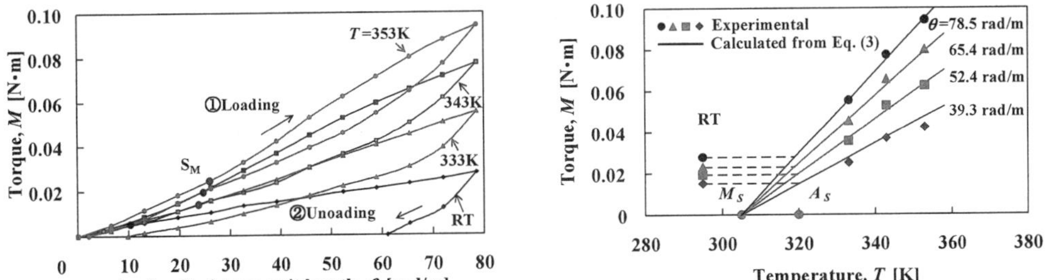

(a) Relation between torque and angle of twist.

(b) Relation between torque and temperature in the loading process.

Fig. 2 Results of torsion test for heat-treated material

$$M = awt^3 G\theta(T - M_s), \tag{3}$$

where the factor a depends on the ratio w/t. The values of G at temperatures above and below A_s differ in a ratio of about 3:1. Taking the average value G = 20 GPa and a = 1.61×10^{-2} K^{-1}, the calculated results can be found as shown by the solid lines in Fig. 2(b). As can be seen, the dependence of M on θ and T is well described by Eq. (3). As expressed in Eq. (3), the effect of thickness t on M is significantly greater than the influence of w.

Estimation of MT starting point. The starting point of the MT can be estimated based on the Mises-yield theory for the MT stress which has been validated for a TiNi SMA from experiments done on a tube under complex loading[6]. The torsional transformation stress τ_M is determined from the MT starting stress σ_M obtained from the tension test, as $\tau_M = \sigma_M / \sqrt{3}$. The torque at the start of the MT is calculated using the torsional stress obtained from the theory of elasticity for a thin strip[5]. The relationship between torque M and the maximum torsional stress τ_{max} in the elastic region of the rectangular-section bar can be expressed as follows

$$M = k_2 wt^2 \tau_{max}, \tag{4}$$

where the coefficient k_2 depends on the ratio of width to thickness w/t and is about 1/3 for w/t>10. Let us assume that the MT starts when τ_{max} becomes τ_M. The start of the MT is determined to be at the point where the calculated value of the torque coincides with the value obtained from the torsion test, as shown by the solid circle at point S_M in Fig. 2(a). As it can be seen, the rate of increase in the torque gradually rises after the MT starting point.

Energy storage and dissipation. The area under the loading curve of the relation between the torque and the angle of twist corresponds to the work done during loading[7]. The area under the unloading curve corresponds to the recoverable strain energy E_r. The area inside the hysteresis loop during loading and unloading corresponds to the dissipated work W_d.

The relations between E_r and W_d and temperature T at θ= 78.5 rad/m (total angle of twist $\phi = \pi$) for both the as-received and heat-treated materials are shown in Fig. 3. It can be seen that E_r increases markedly in proportion to T as the torque during unloading increases with increase of temperature. The relation between E_r and T is expressed by the following equation

$$E_r = b(T - A_s),$$
$$b\text{=1.86 mJ/K, } A_s\text{=295 K : As-received,} \tag{5}$$
$$b\text{=3.68 mJ/K, } A_s\text{=320 K : Heat-treated,}$$

where A_s denotes the reverse transformation starting temperature under no stress. The calculated results represented by the solid lines in Fig. 3 agree well with the experimental ones. Therefore the characteristics of energy storage in a torsion bar are superior at high temperatures in the superelastic region. The relation between E_r and T reflects the fact that the reverse transformation stress, which

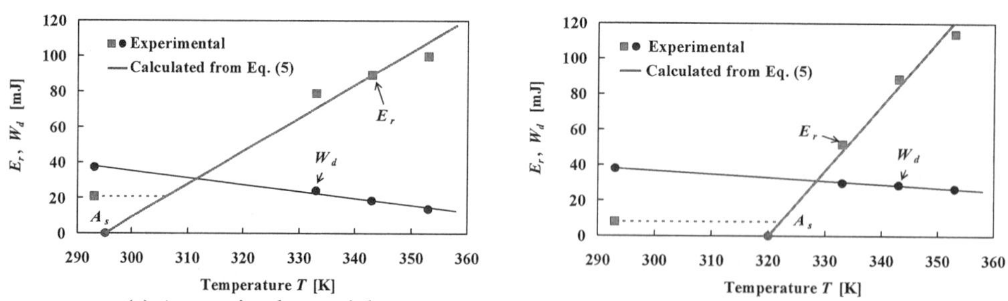

(a) As-received material.

(b) Heat-treated material.

Fig. 3 Dependence of E_r and W_d on temperature at θ =78.5 rad/m (ϕ=π).

acts directly upon E_r, increases proportionally to temperature (T-A_s). E_r increases with an increase of θ. The precise nature of this dependence of E_r on θ is a question for future enquiry.

The dissipated work W_d decreases gradually with an increase of temperature T (Fig. 3). The rate of W_d decrease is small and W_d is only slightly dependent on T. The reason why W_d decreases at a high temperature is because the torques at both the MT starting point and the reverse-transformation finishing point increase with rising temperature. Therefore, the torque range for the M-ϕ hysteresis loop decreases during loading and unloading. The dependence of E_r and W_d on temperature in the thin strip under torsion is similar to that for a TiNi SMA wire under tension[7].

Recovery torque. The relation between recovery torque M and angle of twist θ for the heat-treated material obtained from the recovery torque test is shown in Fig. 4. In Fig. 4, recovery torque M expresses the torque appearing at each temperature T after heating. Recovery torque M_e, which is a cyclic thermal effect, is defined as the difference between the torque appearing at each heating temperature and the torque obtained at room temperature for the given angle of twist.

As it can be seen in Fig. 4(a), the recovery torque increases proportionally to the angle of twist per unit length θ. The recovery torque also increases with an increase of temperature for the same twist angle per unit length θ. As seen in Fig. 4(b), effective recovery torque M_e increases proportionally to temperature rise ΔT. This relationship is approximated by the following equation

$$M_e = c_1\theta(T - A_s), \tag{6}$$

where A_s denotes the reverse transformation starting temperature without stress and A_s = 320 K. The relation between M_e and T is a result of the fact that the recovery stress which directly acts upon M_e increases in the reverse transformation region which is proportional to temperature (T-A_s).

Let us discuss the relationship between effective recovery torque M_e and angle of twist per unit length θ based on the theory of elasticity [5]. The relation between torque M and θ for a rectangular-section bar in torsion is expressed by Eq. (2). Considering the relation between M and θ expressed by Eq. (2), Eq. (6) becomes

$$M_e = cwt^3G\theta(T - A_s), \tag{7}$$

where the coefficient c depends on the ratio w/t. The calculated results where G = 20 GPa and c=1.06×10^{-2} K^{-1} are shown by the solid lines in Fig. 4(b). It can be seen that the overall inclination of M_e is expressed by Eq. (7). However the calculated results do not precisely match the experimental results. This point requires further investigation.

Rotary movement of opening and closing door model. Photographs of the rotary movement of an opening and closing door model using the SMA thin strip (width w=5mm and thickness t=0.25mm) under heating are shown in Fig. 5. The rotational angle of the door ϕ= π/2 given at a low temperature diminishes to ϕ= 0 upon heating.

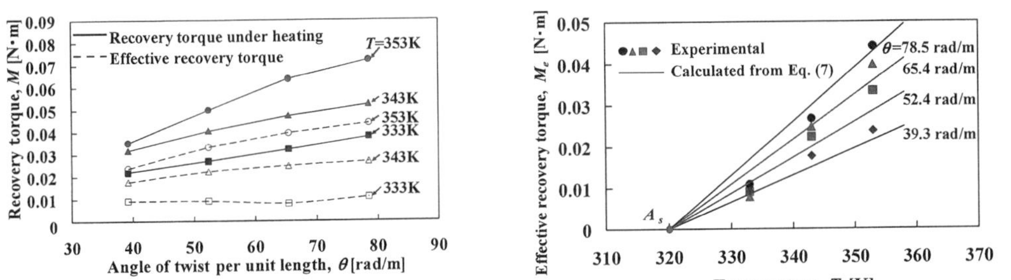

(a) Relation between recovery torque and angle of twist.

(b) Relation between effective recovery torque and temperature.

Fig. 4 Recovery torque for heat-treated material.

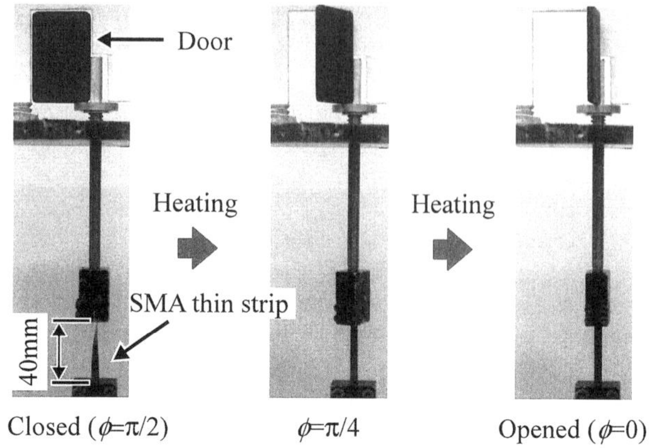

Fig. 5 Rotary movement of a door driven by SMA thin strip under heating.

Thus, if an SMA thin strip is used, a rotary driving element with a small and simple mechanism can be developed. Although this model shows only one-way movement, a two-way element can be developed by using a bias element.

Conclusions

In order to investigate the rotary driving characteristics of an SMA thin strip, the torsional deformation properties of the strip were discussed. The results obtained can be summarized as follows.

(1) Torque increases in proportion to angle of twist and temperature (T-M_s). Although the recoverable strain energy increases in proportion to temperature (T-A_s), the dissipated work decreases slightly with an increase in temperature.

(2) The recovery torque increases in proportion to angle of twist and temperature (T-A_s).

(3) It is confirmed that a rotary driving element with a small and simple mechanism can be developed by using the TiNi SMA thin strip.

References

[1] H. Funakubo (ed): *Shape Memory Alloys* (Gordon and Breach Science Pub., 1987) pp. 1-60.

[2] K. Otsuka and C. M. Wayman (eds): *Shape Memory Materials* (Cambridge University Press, 1998), pp. 1-49.

[3] D. C. Lagoudas (ed): *Shape memory Alloys, Modeling and Engineering Applications* (Springer, 2008) pp. 1-52.

[4] J. H. Mabe, R. T. Ruggeri, E. Rosenzweig and C. J. Yu: Smart Struct. Mater : 2004, Proc. SPIE. **5388** (2004), pp. 95-109.

[5] S. T. Timoshenko and J. N. Goodier: *Theory of Elasticity, 3rd ed.* (McGraw-Hill, 1982), pp. 307-313.

[6] K. Hashimoto, M. Tokuda, T. Inaba, A. Shiode, P. Sittner and B. Bundara: Key Engineering Materials. Vols. **233-236** (2003), pp. 547-552.

[7] E. Pieczyska, S. Gadaj, W. K. Nowacki, K. Hoshio, Y. Makino and H. Tobushi: Science and Technology of Advanced Materials. **6** (2005), pp. 889-894.

Solid State Phenomena Vol. 154 (2009) pp 53-58
online at http://www.scientific.net

Structure and properties of the high temperature nitrided/oxided surface of Ni-Ti alloy

J. Lelatko[1,a], T. Goryczka[1,b], T. Wierzchoń[2,c], H. Morawiec[1,d]

[1]University of Silesia, Institute of Materials Science, 40-007 Katowice, Bankowa 12, Poland

[2]Warsaw University of Technology, Faculty of Materials Science and Engineering, Wołoska 141, 02-507 Warsaw, Poland

[a]jlelatko@us.edu.pl, [b]goryczka@us.edu.pl, [c]twierz@inmat.pw.edu.pl, [d]morawiec@us.edu.pl

Keywords: shape memory; NiTi alloy; nitride-oxide layers

Abstract. The nearly-equiatomic NiTi (50.6 at.%Ni, A_f = 10°C) alloy was nitrided at 1073 K using glow discharge technique. Additionally, at the end of the nitriding some amount of oxygen was added.

Phase identification and thickness of the nitrided/oxided layer was done using electron microscopy method and the X-ray reflectivity measurements. The chemical composition of the layers was determined using energy dispersive spectrometer.

Combination of nitriding and oxidation of the NiTi surface produces nitride/oxide complex layer of a nanocrystalline structure. The average thickness of the obtained layer was about 0,5 μm.

The potentiodynamical studies in Tyrod's solution reveal good corrosion resistance of obtained nitrided/oxidized NiTi surface proving its high quality.

Introduction

The characteristic properties of NiTi alloy like shape memory effect, superelasticity and good biocompatibility, create possibility to use it as a metallic biomaterial for orthopedics, dental application, vascular and organ surgeries [1,2]. Regardless of the fact that most previous studies showed good clinical results of nickel-titanium implants, high nickel content in NiTi alloys and possibility of the nickel ions release limit their medical application for long-term implants. In order to reduce this menace and improve the biological resistance, the surface of the NiTi alloy has been covered by protective layers. Titanium oxide [3,4], titanium nitride [5,6] or diamond-like carbon layers (DLC) [7] have been found as good candidates for layers, which sufficiently protect human body. Additionally, combining the sequence of the layers increases alloy corrosion resistance and biocompatibility [8]. The layers of titanium nitride are characterized by low chemical reactivity, high hardness and resistance to wear [5]. Titanium-oxides, especially TiO_2 [9], increase the stability of the surface layers by protecting the bulk material from corrosion, creating a physical and chemical barrier against nickel release and modifying the oxidation pathways of Ni [10].

Relevant deposition techniques of nitriding such as physical vapor deposition (PVD), chemical vapor depositions (CVD) and ion implantation have been intensively investigated and showed promising results [5,11].

The presented work reports results of obtaining the nitride-oxide layers, which was deposited on the NiTi alloy using the glow discharge nitriding technique.

Experimental procedure

The coating layers were produced on flat specimens of the nearly-equiatomic NiTi shape memory alloy (50.6 at.% Ni), delivered by AMT (Belgium). Such chemical composition cause that, after solution treated at 1073 K for 20 minutes followed by quenching into ice-water, the finish of reverse martensitic transformation is below room temperature (A_f = 283 K). It means that at room temperature this alloy posses the B2 structure and superelastic properties.

Surface of the Ni-Ti flat specimens, in dimension of 14 x 12 x 0.8 mm, were mechanically polished using 0.1 μm diamond suspension to ensure good adhesive properties of the layers. Surface layers were produced by nitriding under the glow discharge technique at 1073 K. The process was carried out for 20 minutes. At the end of the nitriding some amount of oxygen was added. It resulted in coating of the Ni-Ti bulk material by the nitride-oxide layer.

Phase identification of the layers was done using diffraction patterns registered with use of the grazing X-ray diffraction method (GIXD) at constant angle 0,15, 0,3 and 0,8 degrees. Experiments were carried out applying X'Pert Pro diffractometer equipped with Eulerian cradle. Microstructure characterisation of the layer was carried out on a cross section of the specimens using JEOL transmission electron microscope (TEM) JEM 3010 operated at 300 kV and equipped with EDS system. The thin foils from a sample's cross-section were prepared by the ion beam polishing, using the PIPS ion polishing system. The chemical composition of the layers was determined using the energy dispersive spectrometer (EDS) attached to the JEOL scanning electron microscope (SEM) JSM 6480.

The corrosion resistance of the nitride-oxide surface was tested using the cyclic potentiodynamic polarization method. Experiments were carried out in the physiological Tyrod's solution which substituted the body fluid.

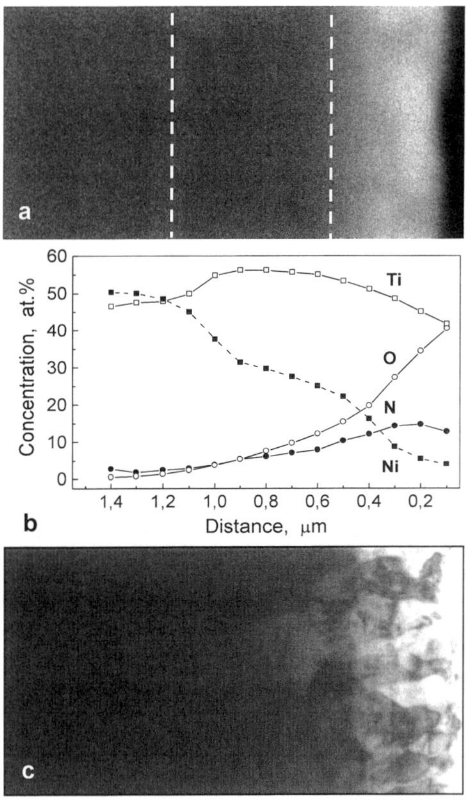

Fig. 1 Cross-section SEM (a) and TEM (c) images of the nitride/oxide layer on the NiTi alloy (a). b – change of the concentration of Ti, Ni, N and O with the depth of the layer.

Results and discussion

Morphology, thickness, structure and quality of the cross-section nitrided/oxidized surface of the NiTi alloy were evaluated from observations using the scanning and transmission electron microscopes. Figure 1 shows SEM and TEM images and the distribution of Ti, Ni, N and O, measured using the EDS technique, on the cross-section of the nitrided/oxidzed at 1073 K Ni-Ti alloy. The observed, average distance of the changes of chemical composition from the top of the surface to inside of the base Ni-Ti alloy was about 1.2 μm. The curve of the analyzed elements distribution shows high amount of oxygen and titanium close to the surface of the layer. At the same point content of nitrogen was about 12 at.%. The increase of the nitrogen content is observed down to the depth of 0.3 μm. Intensive increase of oxygen in regions close to the surface indicates that titanium oxide is created at the top. Simultaneously, beginning at distance of 1μm from the top of the surface continuous decrease of nickel content was noticed.

The surface of the layer practically does not show presence of nickel, while it was stated inside the layer. The nickel content was decreasing with the depth increase. The significant drop of the nickel amount was observed at 0.3 – 0.4 μm below the surface of the layer. Such distribution of the elements confirms that oxygen added at the end of the nitriding process limits the nickel migration to the surface [10].

Change of the contrast on SEM image and the increase of titanium and large decrease of nickel in the area between 1.2 and 0.5 μm below the top of the surface suggest, that during nitrided/oxided process the phase containing higher amount of titanium is created. Such phase was identified during nitriding process using physical, chemical vapor depositions in ion implantation process [5,11]. In a result of titanium diffusion and interaction the Ti_2Ni phase was identified [12].

The structure analysis of the nitride/oxidized surface of the NiTi alloy was carried out using the grazing incident X-ray diffraction method (GIXD) at a constant angle (Fig. 2).

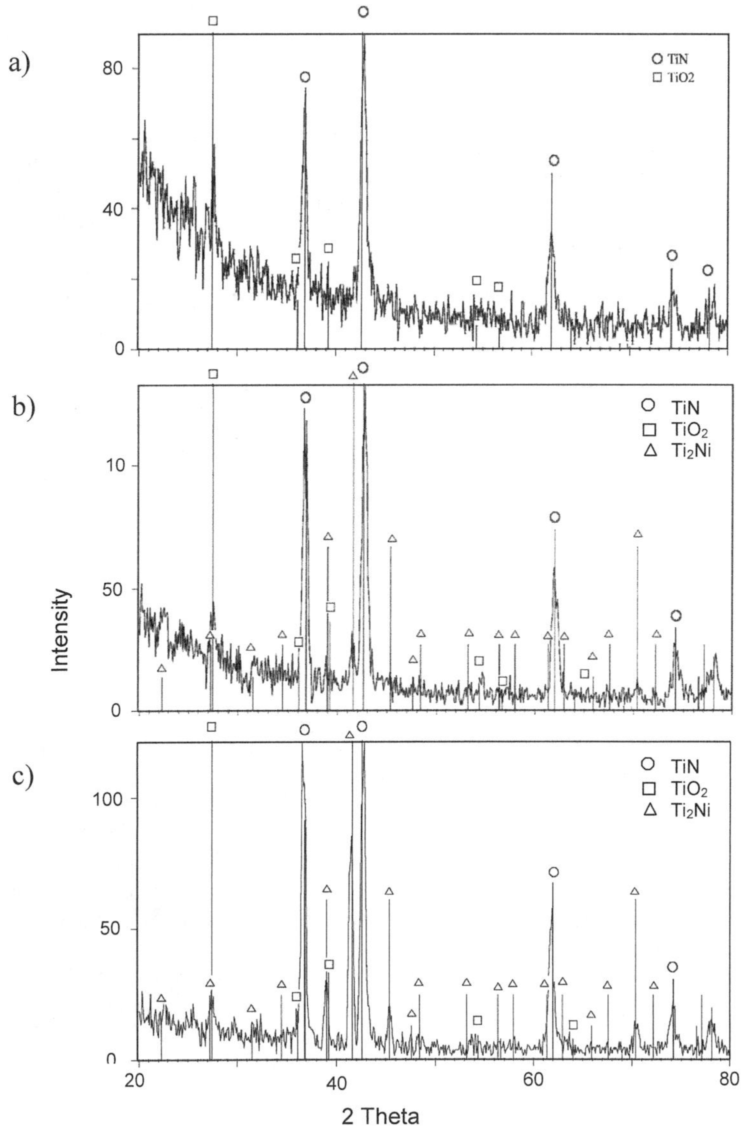

Fig. 2 X-ray diffraction patterns obtained at constant angle 0.15 (a), 0.3 (b) and 0.8 (c) degrees

The diffraction patterns obtained at an angle of 0.15 deg. shows that on the top of surface only the TiN and TiO_2 phase exists. Low intensity of TiO_2 diffraction lines means that volume fraction of this phase is less than TiN phase, i.e. the thickness of TiO_2 layer is rather small.

The increase of the penetration depth (angle between the primary X-Ray beam and sample surface was increased from the 0.3 to 0.8) revealed that the diffraction lines characteristic of the Ti_2Ni phase appeared. The amount of the Ti_2Ni phase increased with the penetration depth of the layer.

The structure of the obtained nitride-oxide nanoscale layer was investigated using transmission electron microscopy. The TEM images (Fig. 3) showed a complex microstructure of the layer. The outer layer, with its thickness of about 0.5 μm possesses nanocrystalline structure, whereas the grains in the intermediate layer reveal a larger size. The grains inside the layer are formed in a shape of columns.

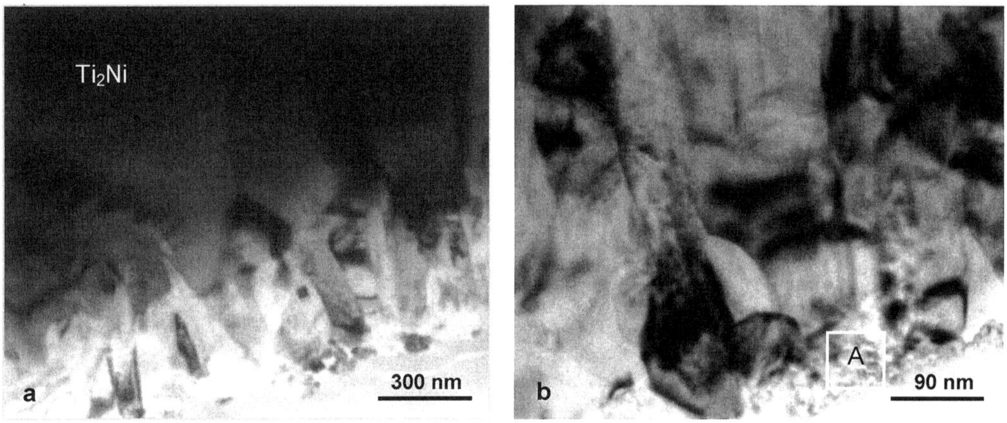

Fig. 3 TEM images of the cross section of the nitrided/oxidized surface layer obtained on NiTi alloy

Transmission electron microscopy investigations confirm the results obtained from the X-ray diffraction method. The phase composition of the layers was identified by indexing the selected area diffraction patterns as well as indexing the FFT patterns obtained from the high resolution images of the nanocrystalline phase. The dark field observation shows that the upper part of the layer is mainly composed from the TiN phase. During the observation carried out using the dark field technique, the grains of titanium oxide TiO_2 were identified. These grains are smaller in size and their volume fraction is relatively small. The grains of titanium oxide are located near the top of the layer. In this part of the nitride-oxide layer, close to the interface of the layer, the areas of structure containing nanograins were observed (area A in the Fig. 3b). The high resolution electron microscopy observation confirmed there, presence of some amount of an amorphous phase between the nanograins of the TiN and TiO_2 phase (Fig. 4).

Low amount of titanium oxide in upper part of the layer and large concentration of oxygen (Fig. 1) suggests that some amount of oxygen creates solid solution with titanium nitride – Ti (N,O).

The nitride-oxide (TiN/TiO_2) layers are homogenous in thickness and grain sizes. Between these layers and NiTi matrix an intermediate layer exists composed from the Ti_2Ni phase (Fig. 3a). Its thickness was about 0.8 μm.

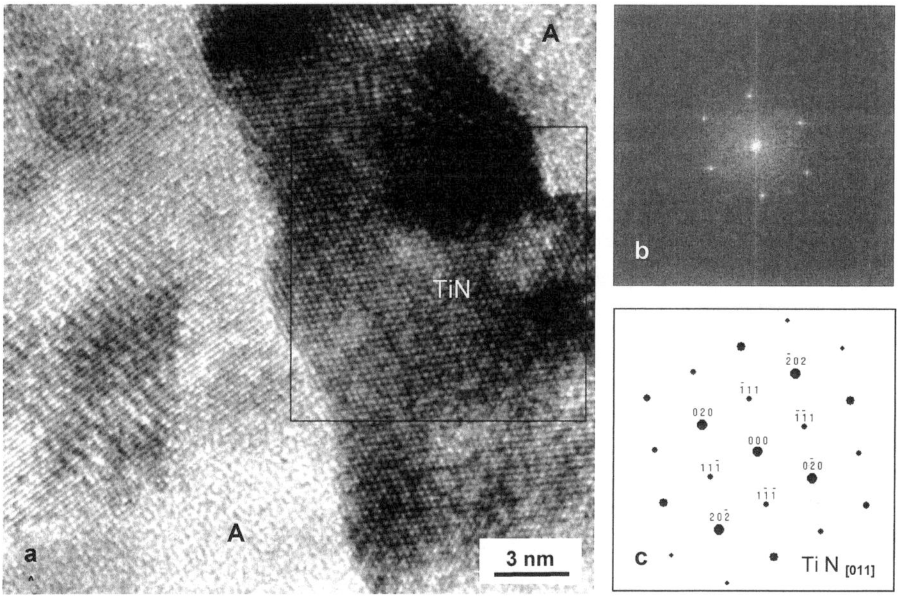

Fig. 4 HREM image of the nitrided/oxidized layer containing areas of amorphous phase A (a), FFT pattern obtained from TiN nanograin (b) and simulated diagram of FFT pattern (c)

The observation of the nitrided/oxidized NiTi shape memory alloy shows that adding some amount of oxygen at the end of the nitriding process changes the morphology of the nitrided surface [6]. At the same conditions of glow discharge (temperature, time) the nitrided surface of NiTi alloy possesses more complex and nanocrystalline structure. In both cases the NiTi alloy is characterised by relatively high value of the corrosion resistance. The potentiodynamical measurements show that the breakdown potential E_{br} of the nitrided/oxidized NiTi alloy is higher than 2000 mV and the corrosion potential $E_{corr} = -0.61$ mV. Thus, the presence of the oxygen on the surface of the layer increases the corrosion resistivity of the NiTi alloys.

Summary

The combination of nitriding and oxidation of the NiTi surface carried out at 1073 K using the glow discharge method, produces layers of a nanocrystalline structure, which consist of the columnar grains of the TiN and TiO_2 phases. Between them small amount of an amorphous phase was identified. The adding of oxygen in the end of nitriding changes the multilayer character of the nitride layer and increases the size of the nanograins of TiN phase. Under the nitride-oxide layer the Ti_2Ni phase was identified. The Ti_2Ni phase creates an intermediate layer between the nitrided/oxidized NiTi surface and NiTi matrix.

Adding of some amount of oxygen at end of the nitriding process decreases the migration of nickel atoms to the surface, which results in biocompatibility improving of the NiTi alloy. On the surface of the layer, practically, oxygen was not confirmed.

High corrosion resistance received after layer deposition proves purposefulness of their application.

Acknowledgements: This work was supported financially by the Ministry of Science and High Education (Project N N507 4587 33)

References

[1] T. Duerig, A. Pelton, D. Stöckel: Mater. Sci. Eng. Vol. A 273-275 (1999), p. 149.

[2] A. Trigwell, G. Selvaduary, in: Proceedings of the 2nd International Conference on Shape Memory and Superelastic Technologies, SMST-97, edited by A. Pelton, D. Hodgson, R.Russel, T. Duering, SMTS Procedings, Pacific Grove, CA, (1997), p. 387.

[4] H. Morawiec, J. Lelątko, G. Stergioudis, T. Goryczka, A. Winiarski, P. Paczkowski: Eng. of Biomaterials Vol. 37 (2004), p. 32.

[5] D. Starosvetsky, I. Gotman: Surf. Coat. Technol. Vol.148 (2001), p. 268.

[6] J. Lelatko, P. Paczkowski, T. Wierzchoń, H. Morawiec: J. Microscopy Vol. 223 (2006), p. 234.

[7] M.G. Fyta, C. Mathioudakis, G. Kopidakis, P.C. Kelires: Thin Solid Films Vol. 473 (2005), p. 56.

[8] T. Goryczka, P. Paczkowski, J. Lelatko, T. Wierzchoń H. Morawiec: Solid State Phenomena Vol.130 (2007), p. 151.

[9] J. Lausmaa, et al.: Mat. Res. Soc. Symp. Proc. Vol. 55 (1986), p. 351.

[10] R.D. Barrett, et al.: Am. J. Orthod. Dentofac. Orthop. Vol.103 (1993), p. 8.

[11] D. Starosvetsky, I. Gotman: Biomaterials Vol. 22 (2001), p. 1853.

[12] S.K. Wu, H.C. Lin, C.L. Chu: Surf. Coat. Technol. Vol. 2 (1997), p. 206.

Solid State Phenomena Vol. 154 (2009) pp 59-64
online at http://www.scientific.net

Structure and Shape Memory Effect in Annealed Ni-Ti-Co Strip Produced by Twin Roll Casting Technique

Tomasz Goryczka

University of Silesia, Institute of Materials Science, Bankowa 12, Katowice 40-007, Poland

goryczka@us.edu.pl

Keywords: shape memory alloy, shape memory effect, martensitic transformation, Ni-Ti-Co, twin roll casting.

Abstract. The ternary $Ni_{47}Ti_{50}Co_3$ alloy was produced using twin rolls casting technique. The strip was annealed at 300^0C, 400^0C or 500^0C for 1 hour. Annealing was carried out in real condition without protection atmosphere. The strips revealed a multi-stage martensitic transformation. However, the origin of this multi-stage transformation is different in the as-cast strip when compare to the thermally treated samples. Thermal treatment increases amount of preferentially oriented grains and in consequence higher degree of the shape recovery can be obtained.

Introduction

Ternary Ni-Ti-Co alloy belongs to a large family, which reveals shape memory effect (SME) [1]. Addition of cobalt causes changes in thermal behavior as well as in sequence of the martensitic transformation [2]. Applying the twin rolls casting as a rapid solidification technique for shape memory alloy (SMA) production significantly influences microstructure and grains formation [3]. Additional thermal treatment opens wide possibility in steering of the transformation behavior [4].

The aim of the present work was to find response of the martensitic transformation and shape memory effect on thermal treatment, which was carried out in real conditions – without protective atmosphere.

Experimental data

Ni-Ti-Co shape memory strip was produced by twin roll casting technique (TRC) using processing parameters reported at [5]. The nominal chemical composition of the alloy was as follows: 50at%Ti, 47at%Ni and 3at% Co. Optimized processing parameters allowed to obtain polycrystalline strips 25 cm long, 45mm wide and 300μm thick. For studies the samples were cut from the central part of the strip and subjected to annealing at 300^0C, 400^0C or 500^0C for 1 hour. Annealing was carried out in real condition without protective atmosphere.

Characteristic temperatures of the martensitic transformation were determined from the curves registered using differential scanning calorimeter (DSC) Perkin-Elmer DSC-7 at a heating/cooling rate of 10deg/min. The temperatures: R_s - start of the R-phase transformation, R_f – finish of the R-phase transformation, M_s – start of the forward martensitic transformation, M_f – finish of the forward martensitic transformation, As – start of the reverse martensitic transformation and A_f – finish of the reverse martensitic transformation were calculated using slope extension method.

Microstructure of the annealed samples was observed applying transmission electron microscope JEM 3010 (JEOL) operating at accelerating voltage of 300kV. For experiment, the thin foils were prepared by the ion beam polishing.

Phase identification was carried out using X-ray diffraction patterns obtained from X'Pert Pro diffractometer. The Philips diffractometer (PW 1130), equipped with the texture goniometer, was used for registering incomplete pole figures.

Shape memory effect was studied by means of elongation measurements versus temperature change under external constant tensile load.

Results and discussion

Martensitic transformation. Influence of the thermal treatment on course of the martensitic transformation was evaluated from the DSC curves. Figure 1 shows comparison of the DSC curves registered for the as-cast strip and annealed at 500^{0}C for 1 hour. In the as-cast strip martensitic transformation reveals its reversibility and occurs in multi-stage (Fig. 1a). During cooling the parent phase B2 transforms to the monoclinic martensite B19' through the R-phase. However, the transformation from the R-phase to the martensite appears in two steps marked as M1 and M2. As it was shown by Morawiec *et al.* [4] both martensites M1 and M2 posses the same crystallographic structure – B19'. The two-stage transformation was a result of dislocations network, which was introduced by deformation. Dislocation network increased the local stress. In order to overcome

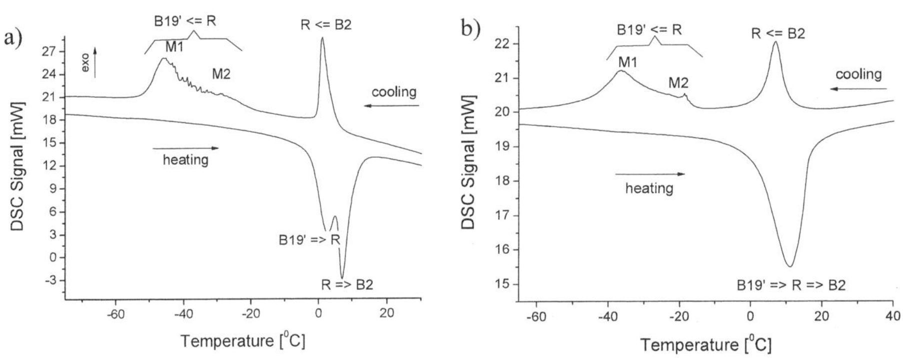

Figure 1. The DSC curves registered for the strip: as-cast (a) and annealed at 500^{0}C for 1hour (b)

the obstacles and to transform the R phase from those regions further temperature lowering was required. Other reason for the two steps martensitic transformation was given by Khalil-Allafi *et al.* [6]. The Author, who studied Ni-Ti alloy after thermal treatment, explained that multi-step transformation is a result of local stress, which appeared from formation of Ni_4Ti_3 precipitates. Still another explanation was delivered by Chrobak *et al.* [7]. The multi-step transformation can be caused by inhomogeneity of chemical composition as a result of the precipitation process. Regions which differ in chemical composition are transformed to the martensite at different temperatures. In consequence, two overlapping thermal peaks can be distinguished on DSC cooling curve.

In the case of the as-cast strip, the local stress appears as a result of structural defects formation originated from the rapid solidification process. Moreover, the local stress is increased by cold rolling, which can be simultaneously realized in the twin roll casting technique. Figure 2a shows a TEM image taken from highly deformed regions in the as-cast strip. Also, presence of primary Ti_2Ni precipitates was stated (Fig. 2b). They are located at grain boundaries. The analysis of chemical composition, carried out at the places close to the grain boundaries revealed comparable content of alloying element to that measured inside of the grain. Goryczka [3] showed that the M2 peak observed on the DSC cooling curve disappeared after 30 thermal cycles and only one step of the martensitic transformation was stated. It was summarized, that the main reason for the multi-step martensitic transformation in the as-cast strip was high density of dislocations.

In general, the annealed strips present similar transformation behavior to that described for the as-cast strip. Figure 1b shows an example of DSC curves measured for the sample annealed at 500^{0}C for 1 hour. Also in this case, the DSC cooling curve reveals a multi-step transformation in the thermal region of formation of the martensite B19'. However, the mechanism of multi-step transition is different then that reported for the as-cast strip.

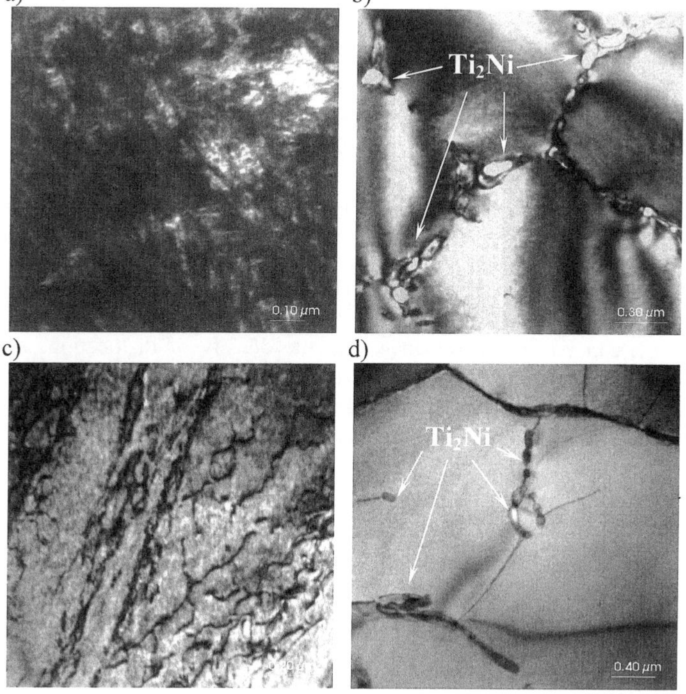

Figure 2. TEM images revealed microstructure of the strip: as-cast (a and b), annealed at 300^{0}C for 1 hour (c) and annealed at 500^{0}C for 1 hour (d).

First, in the result of annealing the deformed regions of the strip recover and the density of dislocation decreases (Fig. 2c). Secondly, the annealing caused precipitation of the Ti_2Ni particles inside the grains. Also the particles were observed at a grain boundaries. Finally, the increase of the annealing temperature, which was carried out without protective atmosphere, intensified the reaction between titanium and oxygen. Phase identification and determination of the surface layers sequence was done using the grazing incidence X-ray diffraction technique (GIXD). The X-ray diffraction patterns were registered at constant incidence angle of 0.3^0, 0.5^0 and 0.8^0. Figure 3 shows the example of the pattern identification done for the strip annealed at 500^{0}C for 1 hour.

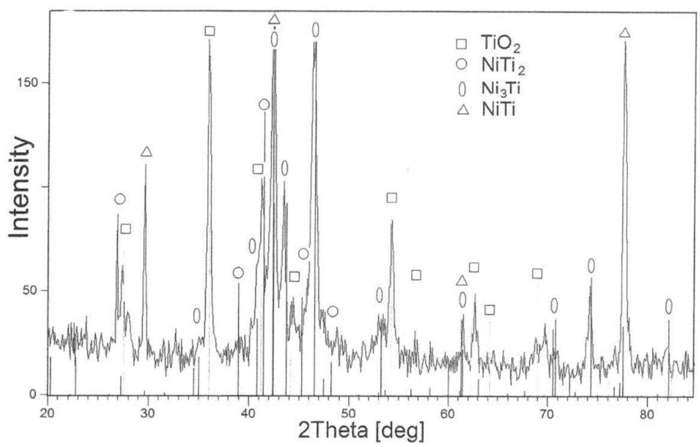

Figure 3. Results of phase identification carried out for X-ray diffraction pattern registered at angle of 0.5^0 the strip annealed at 500^{0}C for 1 hour

In result of oxidation the layer of TiO_2 was formed at the top of surface. Under the TiO_2 layer, mixture of Ni_3Ti and Ti_2Ni phases were found. Diffusion of titanium weakens its amount in Ni-Ti-Co alloy and leaded to inhomogeneity of the chemical composition of regions placed close to the surface of the strip. Considering all these facts it can be concluded that appearance of multi-step martensitic transformation is caused by mechanism described by Chrobak *et al.* [7].

The reverse transformation behavior was observed during heating: the martensite B19' transforms to the R-phase in one stage. In the next step the R-phase is transformed to the parent B2 phase. It is worthy to notice that in the as-cast strip both steps are well separated whereas in the annealed strip they overlapped.

Influence of the annealing on characteristic transformation temperature is shown in figure 4. In comparison to the as-cast strip, all transformation temperatures increased in the samples annealed at 300^0C and 400^0C. It can be due to the facts that the density of dislocations decreased and the annealing conditions (too short time) were not sufficient for introducing inhomogeneity in chemical composition. Significant changes in transformation temperatures were noticed for the sample annealed at 500^0C. All temperatures decreased as a consequence of the nickel content increase in the parent phase.

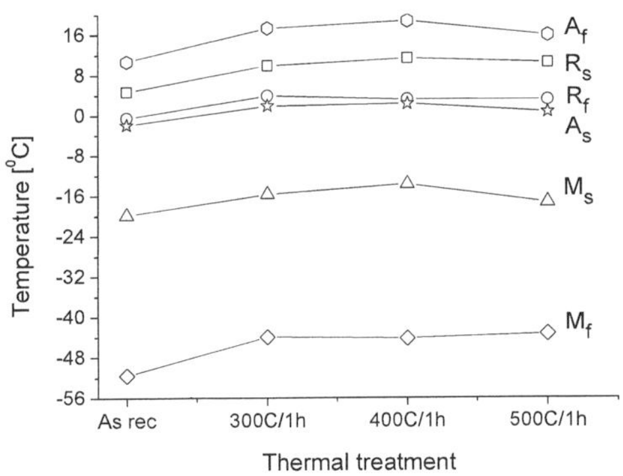

Figure 4. Comparison of the characteristic temperatures of the martensitic transformation for thermally treated strip

Texture. The (110), (100) and (211) pole figures were registered for the as-cast strip as well as for annealed samples. From the measured pole figures the orientation distribution function (ODF) was calculated (Fig. 5) using the computer program LaboTex [8]. In general, the pole densities clustered in similar regions for all the studied samples. Identification revealed that texture consists

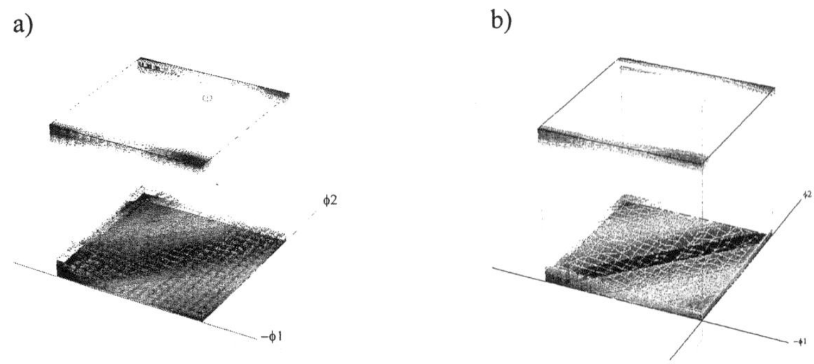

Figure 5. Comparison of ODF calculated for the strip: as-received (a) and annealed at 500^0C for 1 hour (b)

of the axial component <100> and the sheet component {100}<001>. Moreover, calculation of the total amount of grains, which were oriented along the identified texture component, increases for thermally treated samples. For the as-cast strip almost 71% of grains were preferentially oriented whereas this value reaches maximum for the strip annealed at 500^0C and was more than 81% (Fig. 6). It was a consequence of an increase of the textured grains fraction. However, amount of grains oriented along {100}<001> component increased merely of 1.5% when compare the results obtained for the as-cast strip and after annealing at 500^0C. Significant increase of the amount of textured grains oriented along the axial component <100> was noticed – almost 7%.

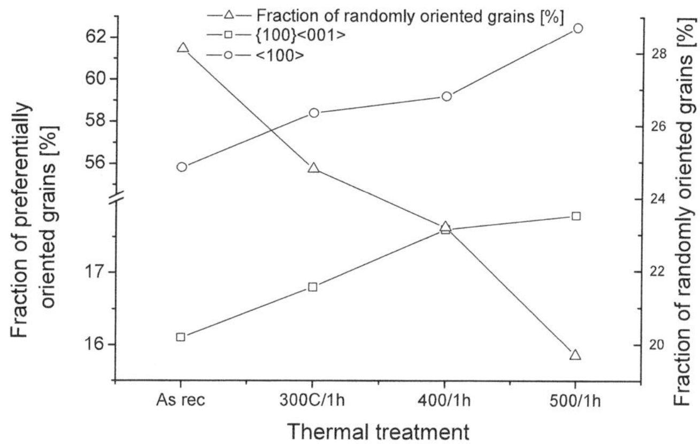

Figure 6. Calculated fraction of grains preferentially oriented along identified textural components

Shape Memory Effect. Figure 7a shows an example of thermal hysteresis obtained from the elongation measurement under external constant tensile load. After completion of each cycle stress was increased of 10MPa and thermal cycling was repeated. From each thermal hysteresis maximal elongation was calculated and compared in figure 7b. It can be clearly seen for the shown

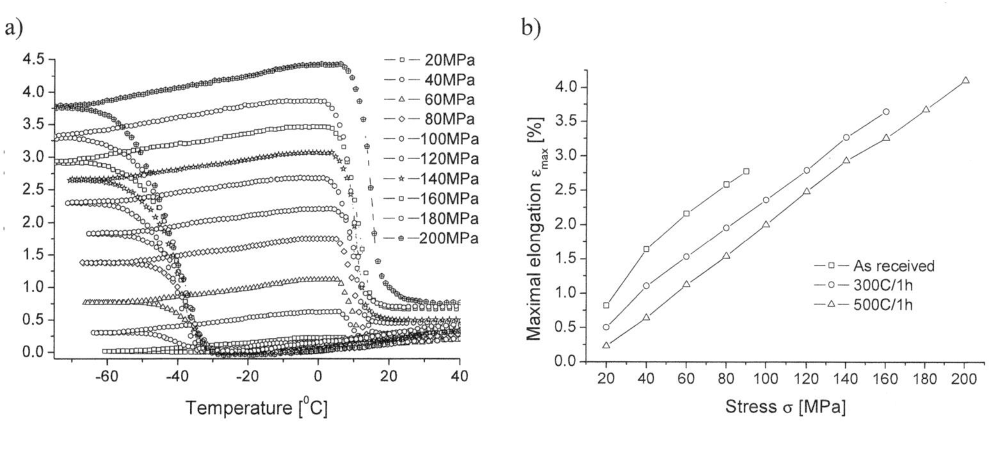

Figure 7. Thermal cycling under constant load carried out for the sample annealed at 500^0C (a) and maximal elongation versus external tensile stress for the strip in as-cast state and thermally treated (b)

example, that for stress higher than 140MPa loops were not closed. In order to close them, additional 3 thermal cycles were carried out. However, when applied stress exceeded 200MPa the plastic deformation appeared. For calculation of the maximal elongation only completely closed loops were taken. The obtained results were set together in figure 7b. For the as-cast strip applying 90MPa allowed to obtain maximal elongation of 2.7%. Annealing at 300^0C caused the increase of the maximal elongation up to 3.6%. In order to obtain this 160MPa was requested. Further increase of the maximal elongation up to 4% was obtained in the sample annealed at 500^0C when stress increased to 200MPa. Such improvement of the shape recovery was possible due to the increase of the volume of the preferentially oriented grains. However, the <100> direction is not privileged one in crystallographic mechanism of martensite formation.

Conclusions

Obtained results allowed drawing the following conclusions:

- Annealing at temperature up to 400^0C causes increase of the characteristic temperatures of the martensitic transformation. In consequence of surface oxidation and increased volume fraction of precipitates, in the sample annealed at 500^0C, the transformation temperatures were lower than in the sample annealed at 300^0C or 400^0C but higher than in the as-cast strip.
- In the as-cast strip and the samples after thermal treatment texture consists of <100> and {100}<001> components. Annealing caused an increase of volume fraction of the preferentially oriented grains.
- Degree of the shape recovery is higher in the annealed strip than in the as-cast one and its value can reach 4%.

Acknowledgement. Studies were financially supported from the project no. N N507 4587 33 funded by the Polish Ministry of Science and High Education.

References

[1] L. Jordan, M. Masse, A. Villafana, G. Bouquet: Proc. of Inter. Conf. Mart. Trans, Monterey, (1992), p. 635

[2] M. Larnicol, R. Portier, P. Ochin: J. Phys IV France 7 (1997), C5-191

[3] T. Goryczka, P. Ochin: J. Mat. Proces. Tech. Vol. 162–163 (2005), p. 178

[4] H. Morawiec, D. Stróż, T. Goryczka, D. Chrobak: Scripta Mat. Vol. 35 (1996), p. 485

[5] T. Goryczka: Mater. Sci. Eng. A Vol. 481–482 (2008), p. 676

[6] J. Khalil-Allafi, G. Eggeler,A. Dlouhy,W.W. Schmahl, Ch. Somsen: Mater. Sci. Eng. A Vol. 378 (2004), p. 148

[7] D. Chrobak, D. Stróż, H. Morawiec, Scripta Mater. Vol. 48 (2003), p. 571

[8] K. Pawlik, P. Ozga; „LaboTex: The Texture Analysis Software", Göttinger Arbeiten zur Geologie und Paläontologie, SB4, (1999).

Solid State Phenomena Vol. 154 (2009) pp 65-70
online at http://www.scientific.net

Performance of shape memory composite with SMA and SMP

H. Tobushi[1, a], S. Hayashi[2, b], Y. Sugimoto[1, c] and K. Date[1, d]

[1]Department of Mechanical Engineering, Aichi Institute of Technology,
1247 Yachigusa, Yakusa-cho, Toyota, 470-0392, Japan.

[2]DiAPLEX Co., Ltd., Ebisu-minami 1-5-2, Shibuya-ku, Tokyo, 150-0022, Japan

[a]tobushi@aitech.ac.jp, [b]hayashi@diaplex.co.jp, [c]pp07711@aitech.ac.jp, [d]pp08715@aitech.ac.jp

Keywords: Composite, Shape Memory Alloy, Shape Memory Polymer, Shape Fixity, Shape Recovery, Recovery Force, Bending

Abstract. The shape-memory composite belt with a TiNi-SMA wire fiber and a polyurethane-SMP sheet matrix was fabricated. The bending actuation characteristics of the belt were investigated by the thermomechanical tests. Residual deflection obtained by cooling under constant maximum deflection is close to the maximum deflection. The residual deflection disappears by heating under no load. Both the rates of shape fixity and shape recovery are close to 100%. Recovery force which appears by heating under constant residual deflection is 93-94% of the maximum force. The model of bending-spring constant of the shape-memory composite belt is proposed and evaluated based on the experimental data.

Introduction

Shape memory materials (SMMs) have been developed as smart materials. In SMMs, shape memory alloy (SMA) and shape memory polymer (SMP) have been used in practical applications[1]. The shape memory effect (SME), superelasticity (SE) and large recovery stress appear in SMA[2],[3]. However, both elastic modulus and yield stress are low at low temperature. This means the fact that rigidity of SMA elements is low at low temperature. The shape fixity (SF) and large shape recovery (SR) appear in SMP[4]-[6]. However, both elastic modulus and yield stress are low at high temperature. This means the fact that rigidity of SMP elements is low at high temperature. If the shape memory composites (SMCs) with SMA and SMP are developed, new and higher functionality of the material can be obtained. That is, shape fixity and high rigity at low temperature and shape recovery and recovery stress at high temperature can be used in the SMC. In the present paper, the characteristics of the SMC belt with SMA and SMP and the bending actuation properties were analysed. A TiNi SMA wire was used for the fiber of the SMC. A polyurethane (PU) SMP sheet was used for the matrix of the SMC. Since the fatigue life of TiNi SMA is very long, it has been most widely used in SMAs. Since PU SMP has various kinds of properties, it has been used in practical applications. The model of bending-spring constant of the SMC belt was proposed.

Characteristics of composite with SMA and SMP

The relationships between elastic modulus and temperature and those between yield stress and temperature for TiNi SMA, PU SMP and mild steel are schematically shown in Fig. 1. As it can be seen, the dependence of elastic modulus and yield stress on temperature is quite different among SMA, SMP and steel. In order to obtain the two-way shape memory effect (TWSME), two-way movement of SMA elements, the combination of SMA and bias materials, for example, steel is used in practical applications[1]-[3]. If the composite materials with these materials are developed, new properties can be achieved. For example, if SMA and SMP are combined, the composite material with large recovery strain, high recovery stress, high rigidity, TWSME and light weight can be obtained. Therefore, if the SMCs with SMA and SMP are developed, new and higher functionality of the material can be expected.

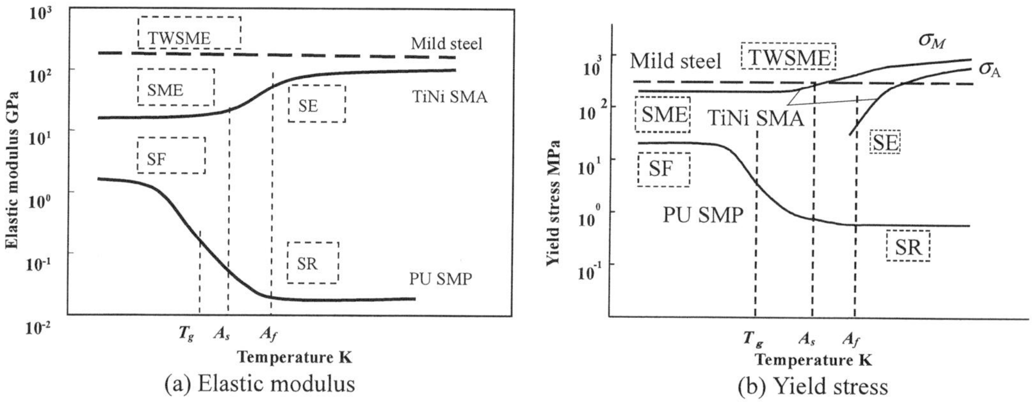

Fig. 1 Dependence of elastic modulus and yield stress on temperature

Materials and Specimen

Polyurethane SMP (Diary MP5510 produced by Mitsubishi Heavy Industries, Ltd.) was used as the matrix of the SMC belt. The glass transition temperature of the SMP sheet T_g was 328K. A TiNi SMA wire (produced by Furukawa Electric Co.) was used for the fibers in the SMC. Diameter and length of the wire were 0.75mm and 50mm, respectively. The finish temperature of reverse transformation A_f of the wire was 325K. The SMC belt was fabricated as follows. First the lower half of the SMP sheet with thickness of 2.1mm was molded by mixing two kinds of liquid in a block. In a few minutes after molding, viscosity of the SMP became high enough to support the SMA wire on the surface of the SMP sheet. SMA wires were put on the lower half of the SMP sheet. The upper half of SMP was molded again on them. The SMC sheet was obtained by curing. Specimens of the SMC belt were cut from the SMC sheet into width of 10mm. The photograph of a specimen is shown in Fig. 2. As can be seen, the SMA wire was located in the central part of the SMC belt. Width, thickness and length of the SMC-belt specimen were 10mm, 4.2mm and 60mm, respectively.

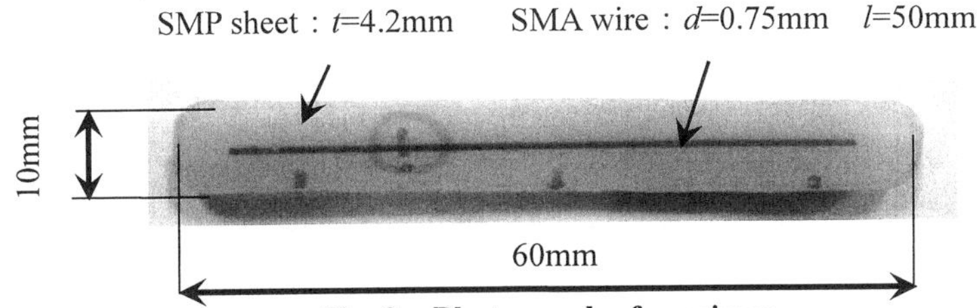

Fig. 2 Photograph of specimen

Evaluation of Elastic Modulus and Spring Constant

If a concentrated force W is applied at the center of a simply supported beam, the maximum deflection y is expressed by the following equation

$$y = \frac{Wl^3}{48EI} \tag{1}$$

where E, I and l denote elastic modulus, second moment of area and distance between two supports. The relationship between W and y is expressed by using the bending-spring constant k as follows

$$W = k \cdot y \tag{2}$$

In the case of a SMA wire of diameter d, by using second moment of area $I_a = \pi d^4/64$, the bending-spring constant k_a is given

$$k_a = \frac{3\pi}{4l^3} E_a \cdot d^4 \tag{3}$$

where E_a denotes elastic modulus of the SMA.

In the case of a SMP beam of a rectangular cross section with width b and height t, by using second moment of area $I_p = bt^3/12$, the bending-spring constant k_p is expressed by the following equation

$$k_p = \frac{4}{l^3} E_p \cdot bt^3 \tag{4}$$

where E_p represents elastic modulus of the SMP.

In the present study, since the volume fraction of the SMA wire in the SMC belt is 1.05%, the bending-spring constant of the SMC belt k_c can be evaluated by a sum of k_a and k_p as follows

$$k_c = k_a + k_p = \frac{1}{l^3}\left(\frac{3}{4}\pi E_a d^4 + 4E_p bt^3\right) \tag{5}$$

With respect to the SMC belt, the bending-spring constant k_c is obtained from a slope of the initial loading curve by using Eq. (2). The SMC belt has a rectangular cross section of width b and height t. Therefore, elastic modulus in bending of the SMC belt E_c is given by the following equation

$$E_c = \frac{k_c}{4b}\left(\frac{l}{t}\right)^3 \tag{6}$$

The experimental and calculated results of k_c and E_c are shown in Table 1. As it can be seen, the bending-spring constant k_c and elastic modulus E_c of the SMC belt can be evaluated by the proposed model. The value of elastic modulus of the SMC belt E_c at T_g+30K is 48MPa which is larger than that for SMP E_p=30MPa at T_g+30K. This means the fact that higher rigidity of SMC elements than one of SMP elements can be obtained in practical applications. As it can be found from Eq. (5), the rigidity of the SMC belt depends strongly on diameter of the SMA fiber and height of the SMP beam.

Table1 Values of elastic modulus and spring constant

Temperature	Experimental value			Calculated value			
	E_p [MPa]	E_a [GPa]	k_c [N/m]	k_p [N/m]	k_a [N/m]	k_c [N/m]	E_c [MPa]
T_g-25K	1150	20	55500	53250	233	53483	1155
T_g+30K	30	70	2240	1389	815	2204	48

Shape Fixity and Shape Recovery

The relationship between force and deflection and that between deflection and temperature obtained from the shape fixity and recovery test by the three-point bending with span l=40mm are shown in Fig.3.

As it can be seen (Fig. 3a), the maximum force appears at deflection of y_{max}=10mm in the loading process at T_g+30K ①. Force decreases almost to zero during the cooling process ② under constant y_{max}. As it can be seen in Fig. 3(b), deflection decreases at temperatures in the vicinity of A_f and T_g during the heating process ④ and approaches to zero at T_g+30K.

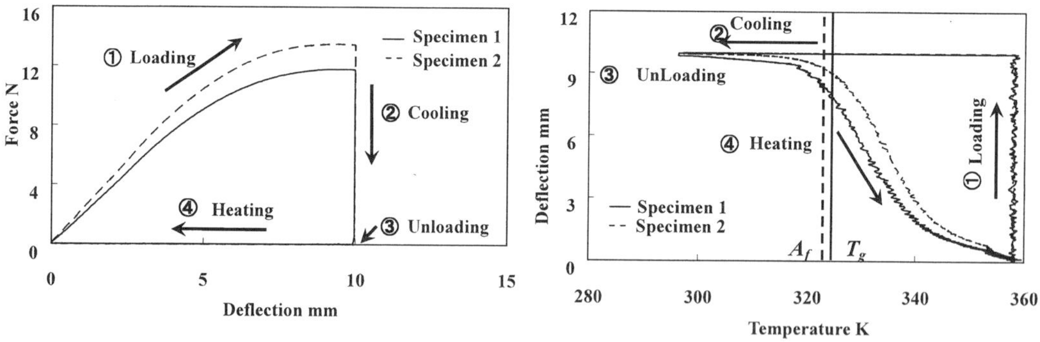

(a) Relationship between force and deflection **(b) Relationship between deflection and temperature**

Fig. 3 Results obtained from shape fixity and recovery test by the three-point bending

Table2 Values of shape fixity and shape recovery rates

	Shape fixity rate R_f [%]	Shape recovery rate R_r [%]
Specimen 1	98.7	98.8
Specimen 2	99.7	99.6

The rate of shape fixity R_f and the rate of shape recovery R_r are defined by the following equations, respectively

$$R_f = \frac{y_u}{y_{\max}} \tag{7}$$

$$R_r = \frac{y_{\max} - y_h}{y_{\max}} \tag{8}$$

where y_u and y_h represent the deflection after unloading at T_g-30K and the deflection after heating up to T_g+30K, respectively. The values of the rate of shape fixity R_f and the rate of shape recovery R_r are shown in Table 2. It can be seen that both R_f and R_r are close to 100%, proving the excellent characteristics of shape fixity and shape recovery. The difference of shape fixity and shape recovery between two specimens is small.

Recovery Force

The relationship between force and deflection and that between force and temperature obtained from the recovery force test by the three-point bending with span l=40mm are shown in Fig. 4.

As it can be seen in Fig. 4(a), the force-deflection curves during the loading ① and cooling ② are almost the same as those for the shape fixity and recovery test ② observed in Fig. 3. Recovery force increases during the heating ④ under constant y_{max}.

As it can be seen in Fig. 4(b), force decreases almost to zero during the cooling ②. In the heating ④, force increases significantly at temperatures in the vicinity of A_f and T_g. The value of maximum recovery force obtained by heating up to Tg+30K is close to the maximum force given in the loading ①. The values of recovery force obtained during heating up to T_g+30K and the rate of recovery force to the maximum force during the loading ① are shown in Table 3. The rate of recovery force is 93-94%. The value of recovery force is higher than 7N which was obtained in the SMP belt without the SMA wire. Therefore, high recovery force can be obtained by SMC elements. The recovery force of SMC elements depends on the volume fraction and the arrangement of SMA fibers.

The development of SMC elements with high functionality is expected by various combinations of SMAs and SMPs. That is, if A_f of SMA is higher than T_g of SMP, the two-way property can be used during heating and cooling.

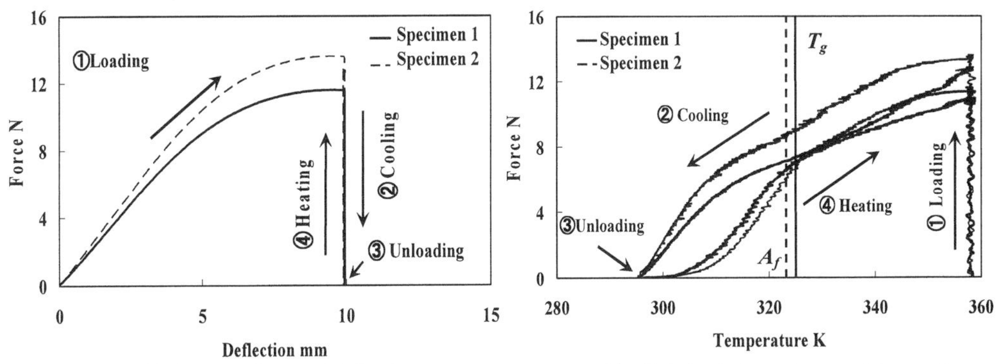

(a) Relationship between force and deflection (b) Relation between force and temperature

Fig. 4 Results obtained from recovery force test by the three-point bending

Table3 Values of maximum force, recovery force and the rate of recovery force to maximum force

	Maximum force [N]	Recovery force [N]	Rate of recovery force [%]
Specimen 1	11.6	10.9	93.2
Specimen 2	13.6	12.8	94.3

CONCLUSION

The SMC belt was fabricated by combining the TiNi-SMA wire fiber and the polyurethane-SMP sheet matrix. The bending actuation characteristics of the SMC belt were investigated. The results obtained can be summarized as follows.

(1) Although spring constant of the SMC belt for bending is large at low temperature, it is small at high temperature. The spring constant of the SMC belt can be evaluated by the proposed model with simple combination of the SMA wire and the SMP sheet.

(2) Force decreases on cooling under constant maximum deflection and high residual deflection close to the maximum deflection is obtained. The residual deflection disappears on heating under no load. The rates of shape fixity and shape recovery close to 100% can be obtained.

(3) The recovery force which appears on heating under constant residual deflection is 93-94% of the maximum force given in the loading process at T_g+30K.

References

[1] M. Mertmann (ed) : *Proc. Int. Conf. on Shape Memory and Superelastic Technologies*, (ASM Inter., 2006), pp. 501-642.

[2] H. Funakubo (ed) : *Shape Memory Alloys*, (Gordon and Breach Science Pub., NewYork. 1987).

[3] K. Otsuka. and C. M. Wayman (ed) : *Shape Memory Materials*, (Cambridge University Press, Cambridge. 1998).

[4] S. Hayashi., "Properties and Applications of Polyurethane Series Shape Memory Polymer", Int. Progr. Urethanes, Vol. 6 (1993): pp. 90-115.

[5] H. Tobushi, T. Hashimoto, N. Ito, S. Hayashi and E. Yamada., "Shape Fixity and Shape Recovery in a Film of Shape Memory Polymer of Polyurethane Series", J. Intell. Mater. Syst. Struct., Vol. 9 (1998): pp. 127-136.

[6] H. Tobushi, D. Shimada, S. Hayashi and M. Endo., "Shape Fixity and Shape Recovery of Polymer Foams", Proc. Instn. Mech. Engrs., Part L.: J. Materials: Design and Applications, Vol. 217 (2003): pp. 135-143.

Solid State Phenomena Vol. 154 (2009) pp 71-76
online at http://www.scientific.net

Shape Memory Assemblies using Ultrasonic Welding

C. M. Craciunescu[1,a], O. Oanca[2,b], D. Dehelean[2,c]

[1] "Politehnica" University of Timisoara, P-ta Victoriei nr. 2, 300006 Timisoara, Romania

[2] National R&D Institute for Welding and Materials Testing, Bd Mihai Viteazul 30, 300222 Timişoara, Romania

[a]craciunescucm@yahoo.com, [b]octavian.oanca@isim.ro, [c]dorin.dehelean@isim.ro

Keywords: shape memory alloys, rapid solidified, welding, ultrasonic energy, welding interface

Abstract The paper describes the research developed in order to assemble shape memory alloys using welding technologies with the aim to reduce the heat affected zone. It is shown that ultrasonic welding can be solutions that could be considered for this purpose. Experiments on welding shape memory alloys have been performed on bulk and rapid solidified ribbons and the interface has been characterized using microstructural observations. The investigations show the difficulties in assembling microfabricated shape memory ribbons with conventional sonotrode/anvil solutions due to the limited thickness of the ribbons and the need to develop microstriations on the surface of the ultrasonic welding tools. It is however considered that ultrasonic welding can be a solution for joining shape memory alloys in similar or dissimilar architectures.

1. Introduction

Shape memory alloys are known to be related to the martensitic phase transition and the behavior is highly sensitive to the chemical composition and thermal history. Based on this known sensibility, it is expected that joining techniques related to welding processes affect the phase transition, thus the shape memory and related properties. Such influences are becoming even more intense when the shape memory alloys dimensions and structural features are in the micrometer range, as is the case for melt-spun ribbons and sputtered films. In particular joining of NiTi shape memory alloys is considered a difficult task, and is an additional factor to be considered when such materials are implemented.

The main difficulties on welding shape memory alloys are related to: overheating phenomena leading to structural changes and local modifications in the chemical composition. Both can lead to degradation or even to a complete loss of the shape memory effect.

Laser welding of Ti [1,2] alloy revealed that the use of high power leads to excessive evaporation during the welding process leading to the formation of brittle phases at the interface which results in interfacial cracks. On the other hand, the analysis of two Ni-Ti parts welded using protective environment showed typical welding areas with melting zone of about 1.5 mm and a heat affected zone of about 3.5 mm [3], with at least part of the shape memory functionality (e.g. pseudoelasticity) affected [4].

For friction and electric resistance welding a high compression force can contribute to intercrystalline cracks closing and to a reduced oxidation during the welding process, while supplementary machining is needed in order to eliminate the extrusions formed [5,6]. Research on Nitinol, suggests that functional properties can be preserved upon friction welding in unprotected atmosphere [7].

Tungsten inert gas (TIG) welding of Ni-Ti showed the need for subsequent heat treatment and training in order to recover the shape memory effect [8]. Welding can also lead to joint degradation as a result of the occurrence of a large heat affected zone, while brazing is another solution used to assemble Ni-Ti materials in dissimilar joints [9]. The structure of the brazing layer depends on the technological parameters used. Increase of power leads to microhardness decrease and increase of the grain size in the heat affected zone, with the shape recovery rate affected by up to 30%.

Shape memory alloys can be welded using various techniques [10]. Comparative studies of various techniques are an effective tool for identifying the best way to assemble particular shape memory alloys. The welding area, the extent of the heat affected zone, compositional and structural changes associated with the welding process contribute to the selection of the best welding technique for an application. All cases shown a 10-20% reduction of the shape recovery rate, more pronounced for TIG welding and less for the laser welding. Apparently, the formation of the martensitic structure is blocked at the limit where dendritic structures form.

The welding process of shape memory alloys is currently a real challenge and the issues become more complex for micro and nanoengineered shape memory alloys. The structural changes related to thermal (and sometimes mechanical) factors may extend to larger surfaces due to the particular low thickness of the shape memory materials. In particular, compositional changes may affect to a great extent the biocompatibility of shape memory alloys due to the potential formation of harmful compounds in the joint. Ultrasonic welding is thus a technique that can significantly reduce the heat affected zone of the welded parts due to the fact that it needs lower joining temperatures and the contact pressure is applied on smaller surfaces. For this reason, it is considered interesting to assess the way this technique can answer current challenges in shape memory alloy joining using microstructural observations of surface and interfacial features.

2. Experimental details

Bulk shape memory alloy ($CuZn_{23}Al_{6}$, $FeMn_{32}Si_{6}$) manufactured by conventional casting methods and ribbons ($Ti_{50}Ni_{25}Cu_{25}$ and $Ti_{50}Ni_{50}$) manufactured by melt-spinning have been ultrasonically welded to similar shape memory alloys or dissimilar bulk materials ($Cu_{99,95}$, $TiAl_{6}V_{4}$). The welding equipment was used in a "weld-to-energy" mode [11]. The power is automatically adjusted during welding (and monitored in % of the maximum ultrasonic power) with an intelligent controller in order to ensure that a proper amount of energy is used for an optimal weld joint.

The sets of materials - according to Table 1- have been welded using the parameters detailed below.

Table 1 Details of the experimental parameters used for ultrasonic welding of shape memory alloys

Set	Materials	Thickness [mm]	Roughness [µm]	Surface quality	Welding position (in contact with: A – anvil; S – sonotrode)	Welding time [ms]	Welding Power [%]	Fs / Pn [N/ bar]	Active surface [mm²]	Specific force [N/ mm²]
1.	$CuZn_{23}Al_{6}$ (ribbon)	0.035	0,4	polished	A	70	75	188 / 1,5	36	13
	$Cu_{99,95}$ (bulk)	0,8	0,4	as spun	S					
2.	$CuZn_{23}Al_{6}$ (bulk)	0,5	0,4	polished	A	20	75	188 / 1,5	14	13
	$CuZn_{23}Al_{6}$ (bulk)	0.5	0,4	polished	S					
3.	$Ti_{50}Ni_{25}Cu_{25}$ (ribbon)	0.032	0,2	as spun	A	3	60	25 / 3,3	9	2,77
	$Ti_{50}Ni_{25}Cu_{25}$ (ribbon)	0.032	0,2	as spun	S					
4.	$Ti_{50}Ni_{25}Cu_{25}$ (ribbon)	0.032	0,2	as spun	S	4	65	27 / 3,5	12	2,25
	$TiAl_{6}V_{4}$ (bulk)	1.5	0,4	polished	A					
5.	$Ni_{50}Ti_{50}$ (ribbon)	0.025	0.2	as spun	A	2	50	22/3,1	9	2.25
	$Ni_{50}Ti_{50}$ (ribbon)	0.025	0.2	as spun	S					
6.	$FeMn_{32}Si_{6}$ (bulk)	0.7	0,4	polished	A	180	75	218/ 1,75	14	15,6
	$FeMn_{32}Si_{6}$ (bulk)	0.7	0,4	polished	S					

The custom-designed welding equipment uses a 3000W piezoceramic converter and has the following characteristics: 20 kHz frequency, 2500 W ultrasonic power, with a 35kVA transformer. The sonotrodes used have been designed for hard and soft materials respectively.

The characteristic elements for the anvil and sonotrode used in the experimental program are shown in Table 2.

Table 2 Profile of the active surfaces of the tools used in the experiments

	Active surface [mm^2]	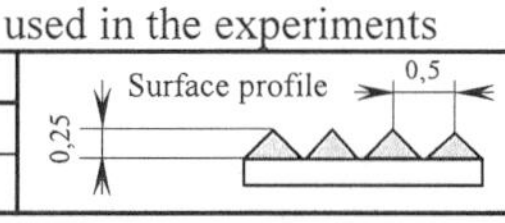
Sonotrode	100	
Anvil	1800	

The welded structures have been characterized using macroscopic examination, X-ray diffraction, optical and electron microscopy, and the results for the welded structures have been compared to the unwelded ones.

3. Results and discussions

Macro structural observations of the joints, exemplified in Fig. 1 show the imprint of the anvil or sonotrode surface profile in the corresponding parts assembled.

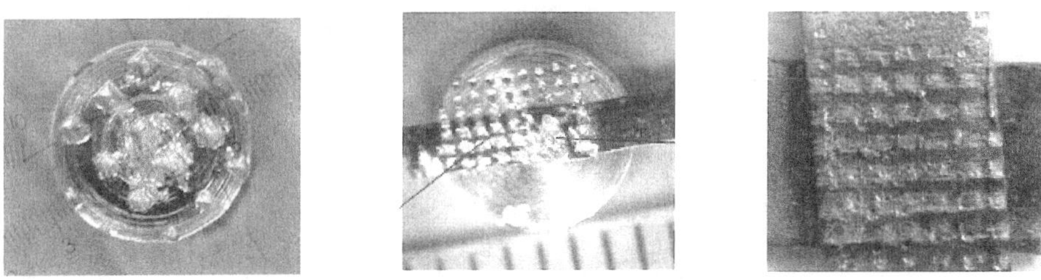

a. bulk $CuZn_{23}Al_6$ on Cu b. $Ti_{50}Ni_{25}Cu_{25}$ ribbon on $TiAl_6V_4$ c. bulk $FeMn_{32}Si_6$ on $FeMn_{32}Si_6$

Fig.1 Surface deformations of the alloy parts in contact with the ultrasonic welding tools.

Local deformations of the surface of bulk shape memory alloys can taken into account prior to welding and can be subsequently eliminated by polishing, however the deformations occurring in thinner structures, such as ribbons are more difficult to eliminate.

Further observations in Fig. 2 detail the local extent of the surface alteration. The imprints frequently show chevrons with orientations related to the direction of the relative movement of the welding tools and the welded parts. The depth of the imprint is influenced by the pressure applied to the active surfaces of the sonotrode and anvil during the ultrasonic welding process.

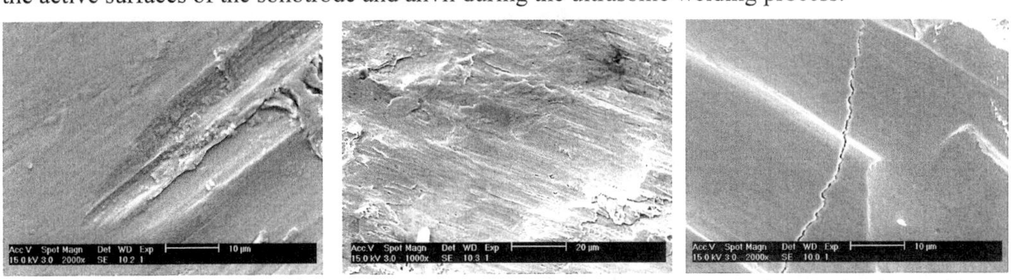

a. Imprint in the bulk $FeMn_{32}Si_6$ surface b. Imprint on the $Ti_{50}Ni_{25}Cu_{25}$ ribbon surface c. Micro cracks observed in $Ni_{50}Ti_{50}$ ribbons

Fig. 2 Details of the surface features for the shape memory alloys ultrasonically welded

Chevrons (fig 2a) or dimples (fig. 2b), depending on the analyzed surface (in contact with anvil or sonotrode or the welded material) have been observed. In addition, microcracks (fig. 2c) may form, parallel with the direction of the relative movement of the welded parts, especially in thinner and more brittle materials, such as ribbons. Such microcracks are initiated at the interface between the welding tool and the material and favored in ribbons by the preexisting microcracks that may occur during the melt-spinning process.

Cross sectional microscopic analysis of the Cu-Zn-Al joints - detailed in Fig. 3 - show the imprint on the outer surface of the welded plates, but since the material is soft even in the thermally aged state, a local deformation at the interface also occurred due to the relative movements of the plates.

Microwelds (Fig. 3 left) have been observed on localized areas and larger welded portions (Fig 3 right) revealing a martensitic structure and precipitates. The surface of the soft Cu-Zn-Al alloy appears to be deformed (with imprints in the range of 100-250 μm height) as result of the pressure applied on the welding tools. The interface also shows deformations (Fig. 3 center) that occurred during the frictional process related to ultrasonic welding.

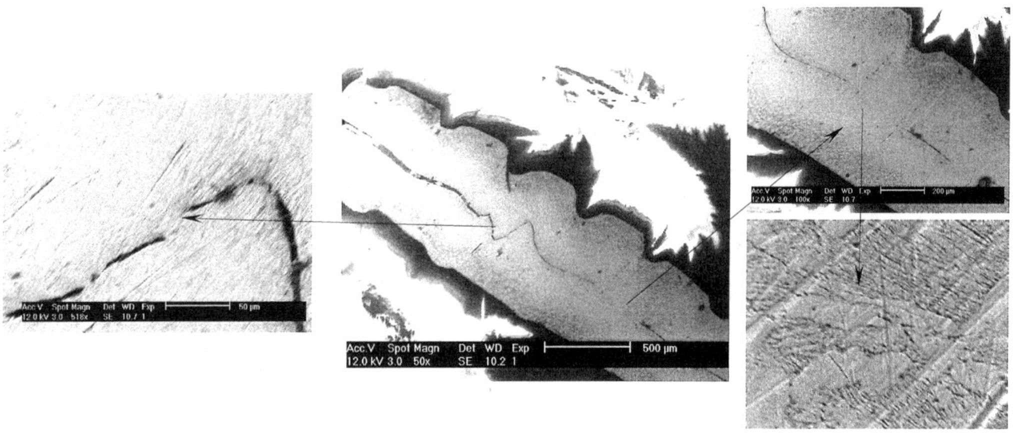

Fig. 3 Cross sections of $CuZn_{23}Al_6$ plates welded using ultrasonic technique

The cross section of the welded interface of the Fe-Mn-Si bulk alloys (Fig. 4) does not reveal a similar influence as for the softer Cu-Zn-Al. The welded interface appears to be solid, as observed by tensile tests and the formation of burrs at the interface similar to the ones observed in friction welding, but of significantly lower amplitude.

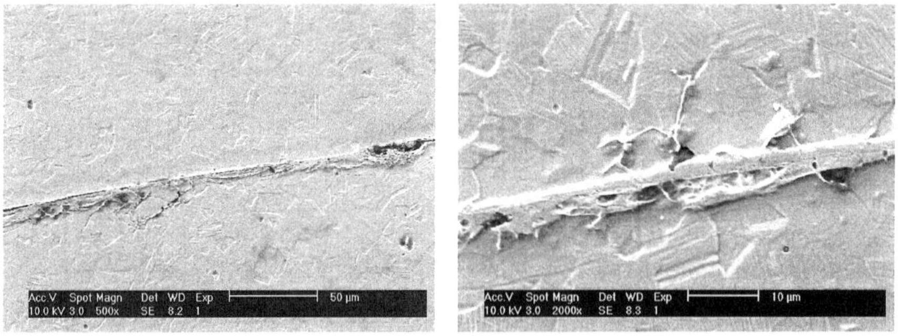

Fig. 4 Details of the welded interface of the $FeMn_{32}Si_6$ alloys

The cross-extent of the interface observed for the bulk Fe-Mn-Si alloys is in the range of 2-4 μms, significantly lower, compared to cases where melting has occurred using other methods. The surface adjacent to the interface does not show signs of deformations, but inclusions and voids present at the contact surface prior to welding may affect the quality of the joint. Compared to joints made out of the same material, welded using electric resistance, where the extent of the heat affected zone is in the range of 2 mm, ultrasonic welding is effective in reducing this area and in localizing it at the interface.

The welding process for ribbons compared to the one for bulk alloys is affected in a higher degree by the surface features of the anvil and sonotrode. The thickness of the ribbon is very small and the tool profile may generate deformation over the entire thickness or even penetrate the ribbon. Figure 5 shows the results of the microscopic observations for $Ti_{50}Ni_{25}Cu_{25}$ ribbons cross welded between each other (Fig. 5a and the corresponding details) and to bulk $TiAl_6V_4$ alloy (Fig. 5b).

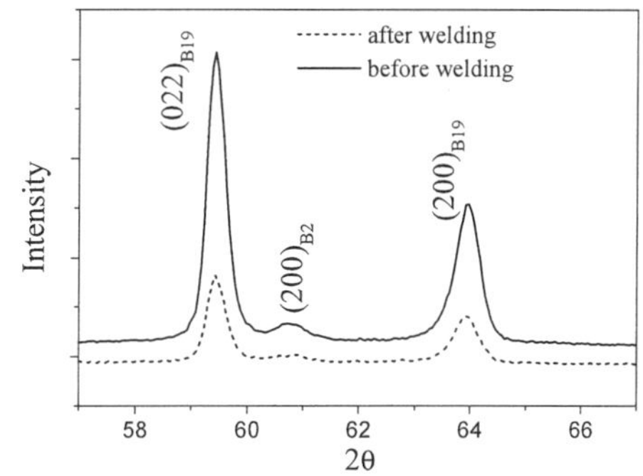

b. X-ray spectra of the ribbons before and after welding (the relative intensity of the peaks in the two graphs is related to the surface exposed)

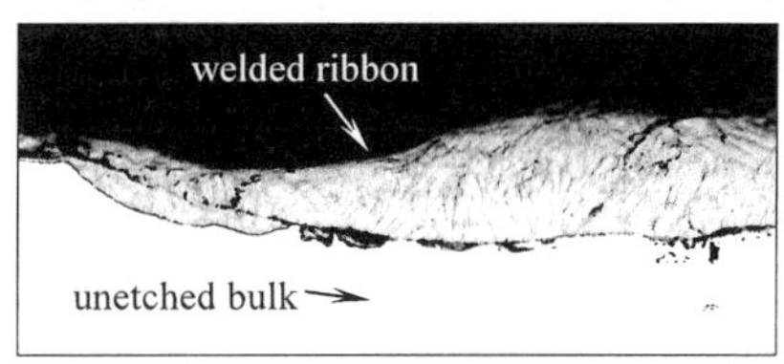

a. ribbons cross welded together

c. ribbon welded on $TiAl_6V_4$ bulk alloy

Fig. 5 Macro and microscopic images of $Ti_{50}Ni_{25}Cu_{25}$ ribbons attached using ultrasonic welding (see text for details)

The ribbons welded using a lower pressure compared to bulk alloys revealed a good interface, while the crack initiated at an angle with respect to the welding interface, as result of peel-off test (bottom detail in Fig. 5a), may suggest a well developed welding interface.

As expected, no major structural changes have been observed by X-ray diffraction, except for textural and grain size changes and optical and electron microscopy did not detect the formation of new structures. The local composition may be affected by impurities at the interface and in case of dissimilar joints one could expect local compositional changes.

For ribbons attached to bulk $TiAl_6V_4$ alloys (Fig. 5c) we observed significant deformations of the ribbon. In addition, cracks were initiated in the ribbons at the surface and at the interface with the bulk alloy, which was also deformed. Further tests are being performed in order to assess the optimal parameters for an improved joint.

Friction welding has been considered as a possibility for joining shape memory alloys even in unprotected atmosphere [7], and the ultrasonic welding appears to lead to observations that localized microfrictional joints can be achieved without major structural influences and with reduced deformations compared to frictional welding.

4. Conclusions

The functionality of shape memory alloys can be less affected if ultrasonic welding is used to assemble the alloys with similar or dissimilar materials, compared to cases where high temperatures are developed during the welding process. The main reason is related to the fact that the ultrasonic energy does not lead to significant heating of the shape memory alloy and the extent of the welding affected zone is limited, since no large scale melting regions are formed.

Our experiments have shown that ultrasonic welding can be used to attach bulk as well as melt-spun ribbons, however, the later case needs special precautions on what concerns surface aspects of the welding tools (sonotrode and anvil). The aspects ratio of the surface profile needs to be adapted to the thickness of the welded materials; otherwise the penetration of the ribbons can lead to permanent damages of the welded areas. Ni-Ti, Fe or copper based alloys can be ultrasonically welded in similar or dissimilar architectures, with areas affected by the process localized around the deformations produced by the contact with the welding tools.

Since it does not require extensive heating, ultrasonic welding can an option to assemble shape memory alloys, especially for applications, where compositional changes that may occur on using conventional welding techniques can affect the biocompatibility.

Acknowledgements

The work presented in this paper has been supported by Romanian National Research Grants. The welding experiments have been made in the "Center of Excellence in Ultrasonic Welding" at ISIM Timisoara. Electron microscopy observations performed at the Institute of Microtechnique of the University of Neuchatel under the grant SCOPES IB 7420-110981/1, financed by Swiss National Science Foundation. The authors gratefully acknowledge this support.

References

1. E.T.F. Chau, C.M. Friend D.M. Allen, J. Hora, J.R. Webster – Mater. Sci. Eng. A (2006).
2. P. Schlossmacher, T. Haas, A. Shussler - J. Phys. IV Coll. C5 (1997) 251.
3. A. Falvo, F.M. Furgiuele, C. Maletta - Mater. Sci. Eng. A 412 (2005) 235–240.
4. A. Tuissi, S. Besseghini, T. Ranucci, F. Squatrito, M. Pozzi - Mater. Sci. Eng. A 273–275 (1999) 813.
5. J. Beyer, P.A. Besselink, J.H. Lindenhovius, in: Y. Chu, T.Y. Hsu, T. Ko (Eds.), Proc. of Int. Symp.on Shape Memory Alloys, China Academic Publishers, Guilin, China, 1986, p. 492.
6. H. Wu Ming, in: Y.Y. Chu, L.C. Zhao (Eds.), Shape Memory Materials and Their Applications, Trans Tech Publ. Inc., Kunming, China, 2001, p. 285.
7. T. Shinoda., T. Tsuchiya , H. Takahashi - Transactions of the Japan Welding Society, Vol.22, No.2(19911001) 102-10.
8. A. Ikai, K. Kimura, H. Tobush - J. Intel. Mater. Syst. Struct. 7 (1996) 646.
9. M.G. Li, D.Q. Sun, X.M. Qiu, D.X. Sun, S.Q. Yin – Mater. Sci. Eng. A 424 (2006) 17–22
10. C. van der Eijk, H. Fostervoll, Z. K. Sallom and O. M. Akselsen - ASM Materials Solutions 2003 Conference, Pittsburgh, Pennsylvania, USA, 13-15 October 2003.
11. G. Flood - Welding Journal, Jan. (1997) 43.

Solid State Phenomena Vol. 154 (2009) pp 77-82
online at http://www.scientific.net

Simultaneous growth of MWCNTs at different temperatures in a variable gradient furnace

V. Grossi [1,a], A. Urbani[1,b], A. Giugni[1,c], C. Cantalini[2,d], S. Santucci[1,e], and M. Passacantando[1,f]

[1]Dipartimento di Fisica, Università degli Studi dell'Aquila, I-67100, Coppito (L'Aquila), Italy

[2]Dipartimento di Chimica, Ing. Chimica e dei Materiali, Università degli Studi dell'Aquila, I-67100, Monteluco di Roio (L'Aquila), Italy

[a]valentina.grossi@aquila.infn.it, [b]ale.urbani@gmail.com, [c]andrea.giugni@aquila.infn.it, [d]canta@ing.univaq.it, [e]sandro.santucci @aquila.infn.it, [f]maurizio.passacantando@aquila.infn.it

Keywords: carbon nanotubes, chemical vapour deposition, scanning electron microscopy, Raman spectroscopy.

Abstract. Multi-walled carbon nanotubes (MWCNTs) have been grown, simultaneously at different temperatures (600 °C, 700 °C, 725 °C, and 750 °C), by thermal chemical vapour deposition (CVD) of acetylene (C_2H_2) gas, in ammonia (NH_3) atmosphere. The samples have been characterized morphologically by scanning electron microscopy (SEM) and a structurally by Raman spectroscopy. These analyses show that, when the temperature of substrate increases, the CNT average diameter and the density distribution decrease, and the CNTs are less defective. However there is a temperature limit (725 °C) for CNT growth. We have grown CNTs onto Si_3N_4/Si substrates with Pt electrodes, in order to realize the gas sensing devices. The CNTs grown at 725 °C, exhibit a fast response and a high sensitivity to NO_2 gas.

Introduction

Due to the exotic electronic structure and to the physical properties multi-walled carbon nanotubes (MWCNTs) promise many applications of great technological importance: conductive wires [1], field emitters [2], new optoelectronic devices [3,4], photosensitive transistors [5], innovative solar cells [6], or chemical sensors capable of detecting a small concentration of molecules, such as CO, NO_2, NH_3 [7].

The commonly used methods for the synthesis of CNTs include arc-discharge [8], laser ablation [9], high-pressure catalytic decomposition of carbon monoxide (HiPCO) [10], and chemical vapour deposition (CVD) [11].

The synthesis of MWCNTs using CVD methods has attracted much attention because of many advantages that high purity, high yield, selective growth, and vertical alignment can be achieved [12]. Therefore CVD is a simplest and economic method to synthesize CNTs. The growth of CNTs could be controlled by varying the growth parameters of CVD process [13] such as furnace temperature [14], reaction time, flow and kind of gas, amount (thickness or concentration) and kind of catalyst [15]. Especially the growth temperature is crucial for selective and controlled growth of CNTs, which is necessary for many applications.

In this work we have grown MWCNTs onto silicon (Si) substrates by thermal CVD of acetylene (C_2H_2) gas in ammonia (NH_3) atmosphere, simultaneously at different temperatures, by using the temperature gradient of the furnace [16,17]. The samples grown simultaneously have been characterized by a scanning electron microscopy (SEM) and a structural characterization has been performed by Raman spectroscopy. We have grown CNTs simultaneously at different temperatures also onto silicon nitride (Si_3N_4) substrates provided with platinum (Pt) patterned electrodes, in view of investigating as resistive gas sensor towards NO_2 for environmental applications.

Experimental

MWCNTs have been grown, simultaneously at different temperatures, by thermal CVD of C_2H_2 gas, in NH_3 atmosphere.

The reactor for the CVD growth is a horizontal quartz tube (100 cm long, 8 cm in diameter) mounted inside a cylindrical furnace 66 cm long (LTF 1200 Lenton). An accurate calibration of the temperature, at various distances from the tube mean position, has been performed by cromel/allumel industrial K-type thermocouple probe inserted into the flow through the exit end of the tube. The schematic diagram of the furnace along with the temperature profile is shown in Fig. 1(b) and (a). The temperature data were recorded when the system reached a steady state in the same conditions of CNT growth.

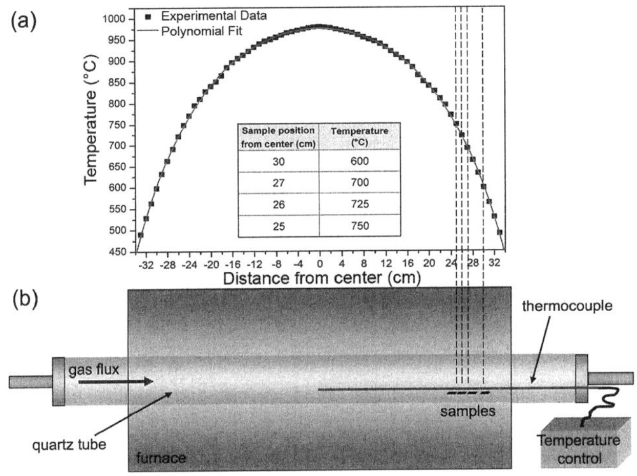

Fig. 1 (a) Calibrated temperature profile across the furnace. (b) Schematic diagram of the furnace. The inset of (a) is a table with the sample positions and the correspondent temperatures.

Si(100) wafers with native layer oxide were used as the substrates. Four substrates, coated with 3 nm of nickel (Ni) film, have been placed onto a quartz holder at different positions (inset Fig. 1(a)) in the reactor and the CNTs have been grown at different temperatures: 750 °C, 725 °C, 700 °C and 600 °C respectively.

The tube has been heated by radiation, in an argon (Ar) flux of 50 sccm, up to 980 °C (temperature in the middle of the reactor) at the heating rate of 20 °C min^{-1}. In order to obtain the Ni nanoclusters, only a 100 sccm NH_3 gas has been introduced into the quartz tube for 20 min at 980 °C. The CNT growth has been performed at 980 °C for 1 hour introducing 20 sccm of C_2H_2 and 100 sccm of NH_3. During the growth the pressure remained stable at ~ 30 Torr. Then the tube has been evacuated and cooled down to room temperature in an Ar flux.

Successively, we have grown CNTs simultaneously at 725 °C and 600 °C on Si_3N_4/Si substrates with Pt patterned electrodes, in view of fabricating a gas detector. The sensitivity of electrical properties to harmful gas NO_2 exposure has been verified. The electrical properties were measured by an automated system. Dry air was mixed by an MKS147 multi gas mass controller with diluted NO_2 mixtures (10 ppm in air) in order to have gas concentrations at the outlet in the range 100 ppb-1.0 ppm. Electrical measurements were carried out selecting the operating temperature of films in the temperature range 25-175 °C. The resistance of the films was measured using a volt-amperometric technique by a Keithley 2001 multimeter.

The morphology of the samples has been examined by a SEM (ZEISS-GEMINI LEO 1530) at 10 kV beam energy. Raman measurements have been performed with a LabRam HR high resolution

Raman microscope (HORIBA-JobinYvon) using a laser with an excitation wavelength of λ_{exc}=632.8 nm.

Results and Discussion

After the nanotubes synthesis a macroscopic inspection of the samples reveals a uniform growth of the CNTs onto Si wafers that are covered by an opaque black film. Investigating the microscopic morphology of the samples with SEM, we observe a dense and intricate net of CNTs grown at 600°C, 700 °C, and 725 °C (Fig. 2(a),(b), and (c) respectively).

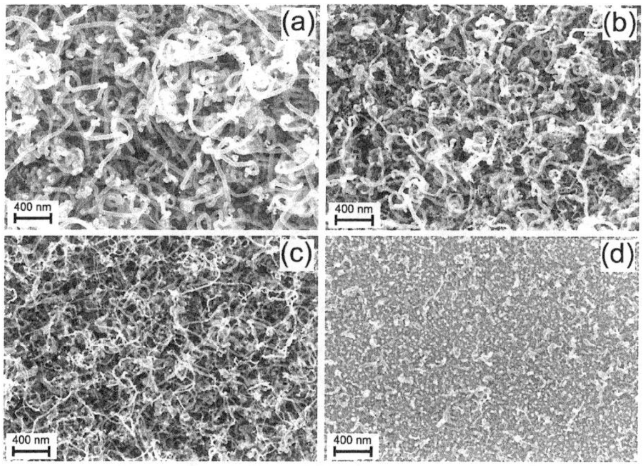

Fig. 2 SEM images of CNTs grown simultaneously at different temperatures: (a) 600 °C, (b) 700°C, (c) 725 °C, and (d) 750 °C.

The Fig. 2(a) shows a SEM image of the uniform and dense distribution of CNTs grown at 600 °C. The CNTs present a quite uniform diameter distribution with an average diameter of 47 nm. The Fig. 2(b) shows a SEM image of CNTs grown at 700 °C. The density distribution does not change significantly with respect to the sample grown at 600 °C. However the size distribution is much larger than the CNTs grown at 600 °C but the average diameter is 23 nm: this is smaller than the CNTs grown at 600 °C, confirming the crucial role of temperature on the CNTs synthesis process. Actually the CNT mean diameter and overall distribution are strongly correlated with the average dimension and density of the Ni nanoparticles onto the supporting substrates. The SEM image in the Fig. 2(c) shows the CNTs grown at 725 °C. They present a uniform and narrow diameter distribution with an average diameter of 17 nm. This means that the MWCNTs grown at higher temperatures have a small number of "walls". The average diameters of all CNT samples were extracted from the SEM images by measuring about 60 structures for each sample. The Fig. 2(d) is a SEM image of CNT grown at 750 °C. As the growth temperature increases up to 750 °C, we obtain only the Ni nanoparticles produced during the CNT growth with few and short CNT fragments, and carbonaceous and graphitic-like contaminants; there is a temperature limit (725 °C) for the CNT growth. During the growth, inside the tube, there is a pressure of ~ 30 Torr. Probably this pressure is too low in order to obtain the conditions for a laminar gas flow along the tube axis. This whirling flow produce the "good sites" (26 and 27 cm from the middle of the furnace), and then "good temperatures", for the CNT growth, where the gas flow direction favors the CNT synthesis.

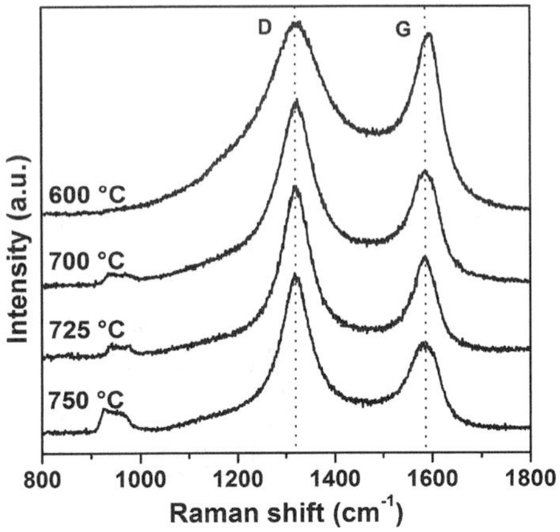

Fig. 3 Raman spectra of CNTs synthesized simultaneously at 600 °C, 700 °C, 725 °C, and 750 °C.

The information about the crystal structure of CNTs and the presence of amorphous material is obtained by Raman spectroscopy characterization [18]. The single-walled carbon nanotube (SWCNT) structure is identified by a peak at ~ 200 cm^{-1} called radial breathing mode (RBM). In our spectra this peak is absent, and then we produce MWCNTs. The first-order Raman spectra of CNTs grown at 600°C, 700°C, 725 °C, and 750 °C are shown in Fig. 3. All spectra show mainly two Raman bands at ~ 1325 cm^{-1} (D band) and at ~ 1580 cm^{-1} (G band). The presence of the D-peak indicates that the CNTs have the structural defects to respect to the high crystalline graphitic layers, indicated by the G-peak. The appearance of the D band can be also interpreted as the formation of nanostructures of amorphous carbon in the samples. In the Raman spectrum of CNTs grown at 600°C the D band appears larger and more structured to respect to the CNTs grown at higher temperatures. The D peak line-width is of ~ 170 cm^{-1} and the ratios of intensities of D and G peaks, I_D/I_G, is 1.05. When the growth temperature increases the D-band line-width decreases, but the intensity ratio of D and G peaks increases. The CNTs grown at 700 °C, 725 °C, and 750 °C have a D peak line-width of ~ 100 cm^{-1}, 70 cm^{-1}, and 80 cm^{-1}, and an I_D/I_G of 1.60, 1.73, and 1.80, respectively. These data confirm that, at higher temperature, we grow CNTs with better graphitic layers but we grow also the amorphous carbonaceous structures with to MWCNTs. When the substrate temperature increases a new peak appears at ~ 950 cm^{-1}: this is a second-order peak of Si crystalline substrate. This means that CNT density distribution decreases at higher temperature growth. In the spectrum of CNTs grown at 600°C, the G peak position is at a higher wave number than that of the G band by ~ 10 cm^{-1}. Actually at ~ 1620 cm^{-1} exists a D' band, that indicates the disorder features of graphitic sheets. At this growth temperature the D' peak is strong and broad, and it is completely overlapping with the G band, resulting in a converged peak at a higher wave number. As the growth temperature increases (700 °C, 725 °C, and 750 °C) D' becomes less intense and narrower and the G peak position does not change noticeably.

The effect of gas environment on the electrical properties of CNTs has been attracted a considerable attention for gas sensing application. Then CNTs have been grown onto Si_3N_4/Si substrates with Pt patterned interdigital electrodes, in view of fabricating a gas sensor. The conditions of CNTs growth on this surface are similar to that of CNTs grown simultaneously onto SiO_2/Si at 600 °C and 725 °C. The CNTs have grown just only on the area between the electrodes, not on the Pt layer. This is due to the diffusion of Ni atoms into Pt film or to the creation of a Pt–Ni alloy. As the Raman spectroscopy indicates, the CNTs have disorder in the graphitic structure. This disorder could induce residual density of states within the band gap giving a predominant metallic character to the CNTs. In order to remove the impurities and amorphous carbon some cycles of temperature have been carried out in dry air from 25 °C to 175 °C and up to 25 °C, by steps of 25 °C. In order to verifier the electrical response of CNT to NO_2, some tests have been carried out exposing the films to dynamic adsorption-desorption cycles (dry air-10 ppm NO_2-dry air) and changing the operating temperature from 100 °C to 200 °C (not shown here). The operating temperature of oxide has been identified to be 150 °C as a trade off between high sensitivity and fast and reproducible base line recovery.

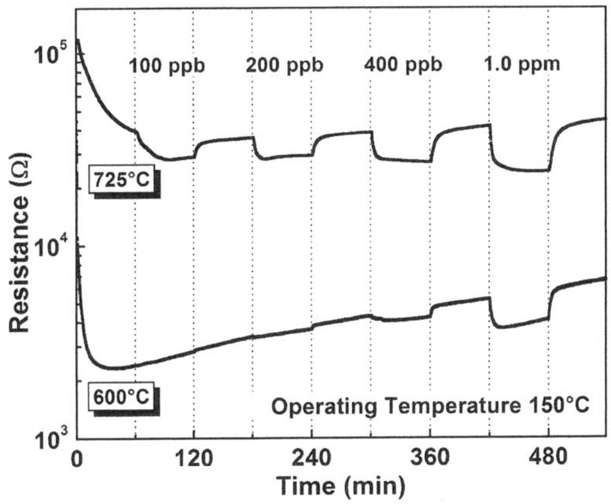

Fig. 4 Resistance changes of the two CNT films grown simultaneously at 600 °C and 725 °C, at an operating temperature of 150 °C and NO_2 concentrations ranging from 100 ppb to 1.0 ppm.

The Fig. 4 shows the sensitivity of CNTs grown at 600 °C and 725 °C to NO_2 gas, at an operating temperature of 150 °C, with various gas concentrations in dry air gas. The sensitivity is defined by the ratio $S = (R_0\text{-}R)/R_0 \times 100$, where R_0 represents the resistance in dry air and R is the resistance in the NO_2 gas environment. The CNTs, grown at 600 °C and 725 °C, have a different order of magnitude of base line resistance: ~ 4×10^4 Ω and ~ 4×10^5 Ω respectively. This is due to the different purity of CNTs as growth and after the annealing treatments. The two samples show a resistance decrease when exposed to NO_2 gas: this change is due to the electrons transfer between the gas molecules and the CNTs. The CNTs grown at 725 °C have a good sensitivity to gas from 100 ppb to 1.0 ppm of NO_2, while the CNTs grown at 600 °C have a noticeable response to gas only from the 400 ppb of NO_2. This last response is less stable and repeatable because the base line has a fast asymptotic raise during the time work. As seen in the Fig. 4, the CNT (725 °C) sensitivity varies from 22.5% to 47.0% when the gas concentration increases from 100 ppb to 1.0 ppm, while the sensitivity of CNTs, grown at 600 °C, is 36.9% at 1.0 ppm. The change of sensitivity from 0% to 22.5% (with 100 ppb) immediately indicates the good sensitivity of CNTs grown at 725 °C to NO_2 gas.

Conclusions

In conclusions, we have grown MWCNTs, simultaneously, but at different temperatures, by thermal CVD. SEM images show a uniform and highly dense distribution of CNTs. As the growth temperature of substrate increases, the CNT average diameter and density distribution decrease. There is a temperature limit for the CNT growth: 725 °C. A structural analysis performed by Raman spectroscopy confirms that the samples grown at higher temperature have less structural defects, less presence of amorphous carbon and lower density distribution of CNTs. By this growth technique we have grown, simultaneously, at different temperatures, CNTs onto Si_3N_4/Si substrates with Pt electrodes, in order to realize the gas sensor devices. Upon exposure to NO_2, at 150 °C operating temperature, we obtain a response with maximum sensitivity (47% at 1.0 ppm of NO_2) for CNTs grown at 725 °C. The electrical resistance of CNTs to oxidizing gases is found to decrease suggesting the occurrence of a p-type response.

Acknowledgments

The authors are grateful to the European Collaborative Project "S-five" for financial support.

References

[1] F. Bussolotti, L. D'Ortenzi, V. Grossi, L. Lozzi, S. Santucci, and M. Passacantando: Phys. Rev. B Vol. 76 (2007), p. 125415
[2] T. Kim, S. Kwon, H. Yoon, C. Chung, and J. Kim: Nanoscience and Nanotechnology Vol. 313-314 (2008), p. 448
[3] M. Passacantando, F. Bussolotti, V. Grossi, S. Santucci, A. Ambrosio, M. Ambrosio, G. Ambrosone, V. Carillo, U. Coscia, P. Maddalena, E. Perillo, and A. Raulo: Appl. Phys. Lett. Vol. 93 (2008), p. 051911
[4] A. Ambrosio, M. Ambrosio, G. Ambrosone, V. Carillo, U. Coscia, V. Grossi, P. Maddalena, M. Passacantando, E. Perillo, A. Raulo, and S. Santucci: Nucl. Instr. and Meth. in Phys. Res. A Vol. 589 (2008), p. 398
[5] J.A. Misewich, R. Martel, Ph. Avouris, J.C. Tsang, S. Heinze, and J. Tersoff: Science Vol. 300 (2003), p. 783
[6] P.V. Kamat: Nanotoday Vol. 1 (2006), p. 20
[7] L. Valentini, C. Cantalini, I. Armentano, J. M. Kenny, L. Lozzi, and S. Santucci: Diamond Relat. Mater. Vol. 13 (2004), p. 1301
[8] D.S. Bethune, C.H. Kiang, M.S. de Vries, G. Gorman, R. Savoy, J. Vazquez, and R. Beyers: Nature Vol. 363 (1993), p. 605
[9] A. Thess, R. Lee, P. Nikolaev, H. Dai, P. Petit, J. Robert, C. Xu, Y.H. Lee, S.G. Kim, A.G. Rinzler, D.T. Colbert, G.E. Scuseria, D. Tomanek, J.E. Fisher, and R.E. Smalley: Science Vol. 273 (1996), p. 483
[10] P. Nikolaev, M.J. Bronikowski, R.K. Bradley, F. Rohmund, D.T. Colbert, K.A. Smith, and R.E. Smalley: Chem. Phys. Lett. Vol. 313 (1999), p. 91
[11] S. Fan, M.G. Chapline, N.R. Franklin, T.W. Tombler, A.M. Cassell, and H. Dai: Science Vol. 283 (1999), p. 512
[12] K. Hata, D.N. Futaba, K. Mizuno, T. Namai, M. Yumura, and S. Iijima: Science Vol. 306 (2004), p. 1362
[13] S. Porro, S. Musso, M. Giorcelli, A. Chiodoni, and A. Tagliaferro: Physica E Vol. 37 (2007), p. 16
[14] C.J. Lee, J. Park, Y. Huh, and J.Y. Lee: Chem. Phys. Lett. Vol. 343 (2001), p. 33
[15] Y. Zhao, D. Huang, and Y. Saito: Nanotechnology Vol. 18 (2007), p. 445608
[16] S. Musso, S. Porro, M. Rovere, Mauro Giorcelli, and A. Tagliaferro: Journal of Crystal Growth Vol. 310 (2008), p. 477
[17] P. Mahanandia and K.K. Nanda: Nanotechnology Vol. 19 (2008), p. 155602
[18] A.G. Souza Filho, A. Jorio, Ge G. Samsonidze, G. Dresselhaus, R. Saito, and M.S. Dresselhaus: Nanotechnology Vol. 14 (2003), p. 1130

Solid State Phenomena Vol. 154 (2009) pp 83-88
online at http://www.scientific.net

Suppressed Field Emission Screening Effect and Electric Field Simulation of Carbon Nanotube-Based Triode Field Emitters

Koohsorkhi, Javad[1, a], Davoudzadeh, Nima[1, b], Mohajerzadeh, Shamsoddin[1,c*], Asl Soleimani, Ebrahin[1,d], Ghafouri Fard, Hasan[2,e]

[1] Thin Film and Nano-electronics Lab., School of Electrical and Computer Engineering, University of Tehran, Tehran, Iran

[2] Electrical Engineering Department, Amir Kabir University of Technology, Tehran, Iran

[a]j.koohsorkhi@ece.ut.ac.ir, [b]nidavidson@gmail.com, [c]mohajer@ut.ac.ir, [d]soleimni@ut.ac.ir, [e]ghafoori@aut.ac.ir

Keywords: Nanotube, Field emission, screening effect, field enhancement factor.

Abstract. In this paper the electric field simulation of carbon nanotubes triode field emission devices is investigated. The geometry of CNTs and their arrangement on the substrate affect the screening effect and filed enhancement factor (β). In an array of nanotubes, for a distance smaller than the CNT length the screening effect is observed. The emission current density is related to the gate aperture and CNT shape. In a triode structure the screening effect occurs in the intertube distances about half of the CNT's length due to the presence of the gate plate. Furthermore, the presence of the metal-oxide gate has a favorable effect to minimize the screening effect and allows one to increase the density of CNT's in each array spot and to increase the cluster array density on the substrate, hence to increase the density of emission current and emitter sites where a high current density is required.

Introduction

Multi-wall carbon Nanotubes (MW-CNTs) are promising for future electronic applications such as field emission devices, owing to their exceptional electronic properties and one dimensional structure. Field emission is the extraction of electrons from the tip of CNT by tunneling through the surface potential barrier under a strong electric field. CNT-based field emission vacuum triode and transistors have numerous potential applications in vacuum microelectronics. They can be used as electronic switch, signal or power amplifier, oscillator, modulator, and control devices completely analogous to conventional solid state p-n diodes and bipolar or MOSFET transistors [1,2]. However, the CNT-based field emission vacuum devices have several advantages over solid-state semiconductor devices.

Field emission vacuum devices can operate at a much higher speed than the p-n diode and transistor because electron transport in vacuum has a higher speed than it does in solid state materials. Consequently vacuum electronic power tubes have in general higher power operation and utilize greater currents than solid-state amplifiers, making them suitable for high frequency applications such as microwave power generation and plasma electronics [3,4]. The vacuum devices can also be used for applications that require electron guns such as electron microscopes (SEM, TEM), electron beam lithography, and Auger electron spectroscopy (AES) [5,6].

Comparing with the other field emitters, CNT-based electron sources revealed an efficient performance at modest vacuum. Field emission flat panel displays (FED's) is another potential application of field emission diode and triode structures [7]. In addition to field emission property, nanotubes posses high electron mobility (7900 cm^2/v.s), high conductance (1 μΩ.cm), low dimension, reduced crosstalk, low weight and high mechanical durability, the ability of high conductivity 10^9A/cm^2 and many other advantages, making them suitable for the fabrication of nano-scale devices [8]. CNT's have shown to have electron emission at field values lower than 1 V/μm and high current densities of over 1 A/cm^2. Detailed reviews of the properties of CNTs as field emitters can be found in the literature [9].

In this paper, we have attempted to investigate the field performance of nanotube emitters and apply it to a novel self-defined structure for the realization of field emission transistors which have been recently reported elsewhere [10]. To investigate the field properties of this configuration, arrays of nanotubes were studied with different inter-tubes distances. Computer simulation has been performed using a CST (Computer Simulation Technology) package. Scanning electron microscopy has been performed using a Cam-scan SEM unit operating at a high voltage of 25kV.

Device Definition

The structure of carbon nanotubes triode field emission devices consists of vertically aligned structures on a silicon substrate which are encapsulated by a metal and insulator bi-layer. The presence of small holes in the metal-oxide bi-layer at the place of a cluster of carbon nanotube resembles a metallic surface on top of the silicon substrate acting as a gate-plate. The formation of field-emission requires a second substrate which is located 100 μm away from the bottom one. Fig. 1 shows the schematic diagram of device.

Aligned CNT's have been made by a dc plasma-enhanced chemical vapor deposition (PECVD) technique to realize multi-wall nanotubes. For this experiment (100) *N*-type silicon substrates are cleaned using RCA #1 solution ($NH_3/H_2O_2/H_2O$ mixture with 1:1:5 ratios) and blow dried with air. The cleaned samples are then placed in a vacuum coating system to deposit a 5–10 nm thick layer of nickel as the seed for the subsequent steps using e-beam evaporation. After unloading the samples, they are placed in a dc-PECVD reactor to grow CNT's. Just prior to the growth, samples are exposed to a 30 sccm flow of hydrogen for a 15 min pre-growth annealing step. This step is performed in the presence of DC plasma with a power density of 4.5 W/cm^2 and at a pressure and temperature of 1.6 torr and 650 °C, respectively.

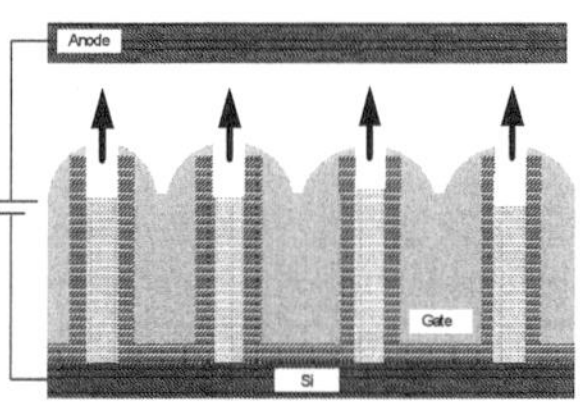

Figure 1: The diagram representing self-defined field-emission devices. The gate, placed around the cathode tube, encapsulates the nanotubes, and the electric field between the gate and cathode controls the emission current which electron emission from the carbon nano-structures is controlled through the surrounding gate.

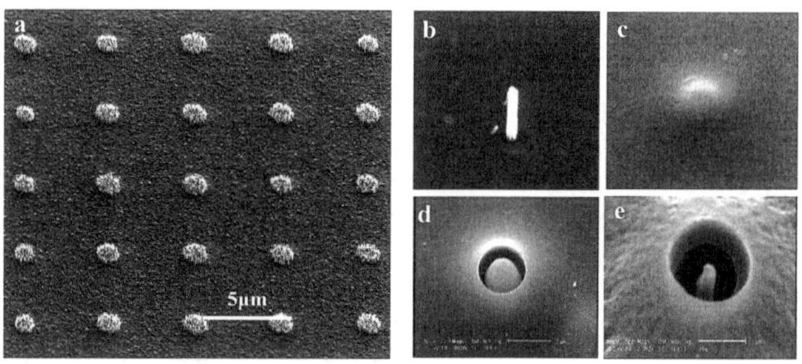

Figure 2: SEM images corresponding to the fabrication process (a): Growth of CNT's on Si substrate using the PECVD method in a cubic shape due to intentional patterning of Ni prior to the growth. (b-d): SEM image of fabrication process, oxide-metal bi-layer coating on the CNT's. (e) SEM image of complete structure of CNT-Based Triode Field Emitters.

Fig. 2a shows a scanning electron microscope (SEM) image of patterned CNT's on a silicon substrate, evidencing a vertical and well-aligned growth. The length of CNTs depends on the duration of the growth step. The well-defined border has been obtained by microlithography of nickel seed layer prior to the growth of CNT's (Fig 2a). Figures 2.b to 2.e collect some of the SEM images corresponding to the fabrication process of CNT including deposition of TiO_2 and Cr(700nm) bi-layer around the CNT's. Titanium-oxide has been used as the insulating material and its deposition occurs at a temperature of 220 °C using an atmospheric pressure chemical vapor deposition technique [11]. The metallic gate (Cr layer) is coated by electron beam evaporation.

Fig 2.c shows the top view of the sample that is coated by a photo-resist layer. By applying the proper voltage to the sample and the top plate the electrons pass through the photo-resist layer, in this condition the bombarded photo-resist would remove from the tip of the coated CNT. Fig. 2d shows the SEM image of the sample after developing the coated CNT by the oxide-metal bi-layers which appears in the center of the gate opening. Fig 2.e shows the triode structure of the filed emitters where each carbon nanotube is surrounded with a metal gate around it. Details about the fabrication process have been previously discussed in ref. [10,12]. By applying a proper voltage between the gate and cathode electrodes, an electron beam is emitted from the CNT tip. Fig. 1 shows the schematic operation of the device and how voltage is applied between cathode and anode electrodes to make the electron emission from the tip of the nanotubes. To apply the proper voltage on the gate we use a photolithography technique to pattern the metallic gate. During this study the distance and voltage between the anode-cathode electrodes are maintained at 50μm and 50 V, respectively. By applying a proper voltage on the gate of the triode structure, a significant control on the emission current is possible. The dimension of nanotubes, inter-tube distance, diameter of the gate openings, distance between nanotubes tip and gate opening are important factors which influence the current emission density.

Results and Discussions

CNT emitters can be fabricated in a variety of configurations depending on the geometry of the nanotubes, their microscopic arrangement, the fabrication process, and the desired emitter structure. The electrical operation of the device is significantly affected by such structural parameters. In a field emission device, the Fowler-Nordheim equation yields the current density as a function of the distance and the applied macroscopic field. Following the work of N. De Jonge, the anode current can be expressed as a function of the anode-cathode voltage as [13].

$$I = AV^{2} \exp\left(\frac{b\phi^{\frac{3}{2}}}{\beta V} \right). \qquad (1)$$

Where I is the current, A and b are constants, Φ is the work function, V the applied voltage between anode-cathode electrodes, and β is the geometric field enhancement factor. The distance between the anode and the CNT tip (substrate) and β are important parameters in FE analysis. Furthermore, the CNT emitter structure affects the geometric field enhancement factor β. Also a variation in the tube tip apex changes the magnitude of the field enhancement factor. So in general the on-voltage for SWCNT emitter is smaller that of MWCNT emitter.

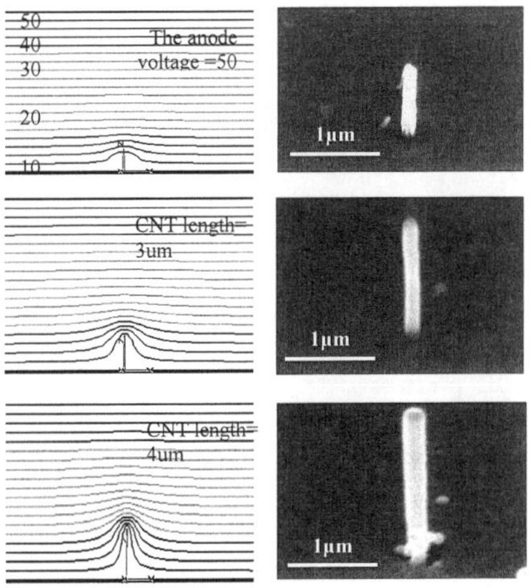

Figure 3: Left: Simulation results of the equipotential lines of the electrostatic field for the CNT with difference lengths of 2-3-4 μm, CNT diameter is 80 nm. The effect of length of CNT in field enhancement factor is clear. Right: SEM images corresponding to the simulation conditions.

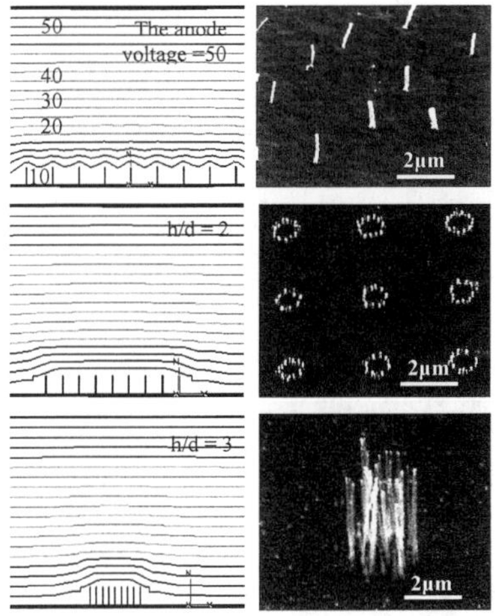

Figure 4: Left: Simulation results of the equipotential lines of the electrostatic field for an array of CNTs with a constant diameter of 60 nm and difference intertube distances h/d (height/distance) =1-2-3 μm, respectively. The screening effect occurs for the samples with distances smaller than the CNT length. Right: SEM images corresponding to the simulation conditions.

At first, we have studied the effect of the length of nanotubes on the enhancement factor which results in the change of emission current from the tip of the nanotubes. The enhancement factor is one of the important parameters which affect the emission current and depends on the shape and the geometry of the nanotubes. The field enhancement factor (β) for the individual CNTs with 50 nm diameter and about 2 μm height is estimated from the Bonard formulae [14] to be 250. As it appears from the left image of Fig. 3, longer nanotubes result in stronger enhancement factor and deform the electrical field more pronouncedly. In the case of a cluster of CNTs, however, longer nanotubes lead to more screening effect. Fig. 3 (right) shows the SEM image of individual CNTs corresponding to the simulation results. The screening effect is an unfavorable effect for the field emission applications and it is more serious for the clustered configurations. To investigate this screening effect of the clustered structure, a collection of 9-by-9 array has been used. The length of CNTs is considered to be 2μm. As depicted in Fig. 4 the screening effect is observable when the spacing between CNTs is shorter than the length of the nanotubes. To avoid this effect, nanotubes in a cluster should have sufficient spacing with each other. All CNT emitters start to emit at relatively low fields (1–2 V/μm). The anode-cathode distance and the voltage are maintained at 50μm and 50 volts during this simulation, respectively. As it is observed from these figures (3 and 4) the best condition for the emission is obtained when the distance between nanotubes is more than the length of each nanotube. It must be born in mind that the above results have been obtained with bare carbon nanotubes and with the absence of a gate plate. These results agree with the results of reference [15,16] where nanotubes are exposed to the anode voltage without a metallic gate.

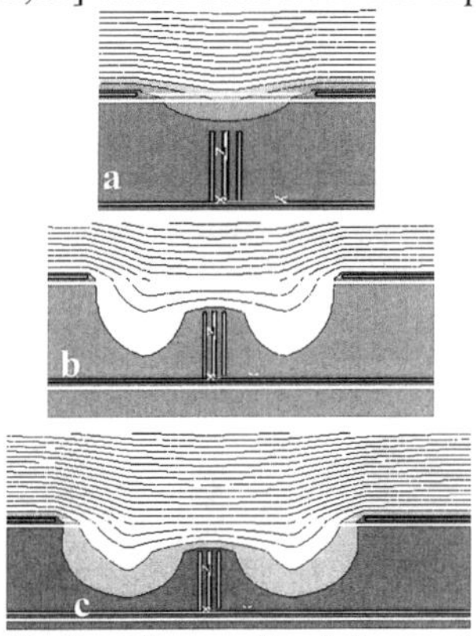

Figure 5: Simulation results of the anode field penetration through the gate opening for the 3-by-3 CNT array. The gate opening diameter is set at 1, 1.5 and 2 μm, respectively. Larger opening diameters lead to a higher penetration of the anode field.

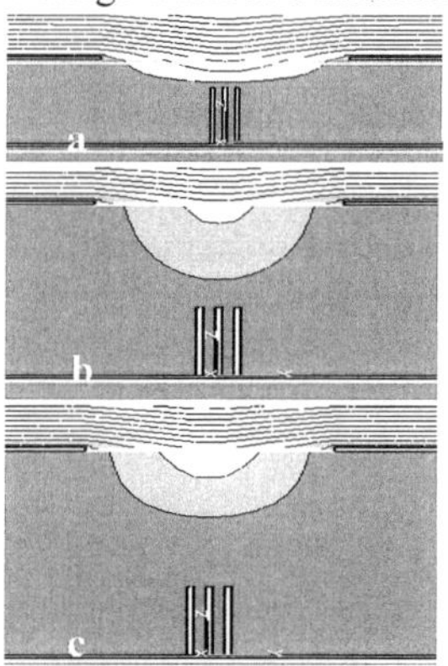

Figure 6: Simulation results for effect of distance between CNT tip and gate opening on filed penetration and FE operation. The simulation made for the spacing of 1-2-3 and 4 μm. The length of CNT's is maintained at 2 μm. FE has been enhanced in the smaller spacing.

To observe the effects of a metal gate around nanotube, a triode structure is investigated. The presence of a surrounding gate significantly alters the operation of the triode devices. In the simulation study, the gate is considered to be a metal plate which has openings on top of the nanotubes. The distance of these openings is proportional to the distances of the clusters on the substrates. Applying gate voltage results in the change of the emission current more significantly and the screening effect is dramatically suppressed. The spacing between the tip of each nanotube and the top gate and also the diameter of the gate openings are important parameters in such structures. To observe the effect of the diameter of the gate openings, we have presented the results of Fig. 5 and 6. As seen from these figures, by increasing the diameter of the opening the electric field due to the anode penetrates through the opening and the resulted field at the tip of the CNT increases. Also if the distance between the gate plate and the CNT tips is reduced, the electric field from the anode electrodes reaches the tip of the CNT more easily. Based on this investigation, the proper diameter for the opening is about the distance between the tip of the nanotube and the opening of the gate.

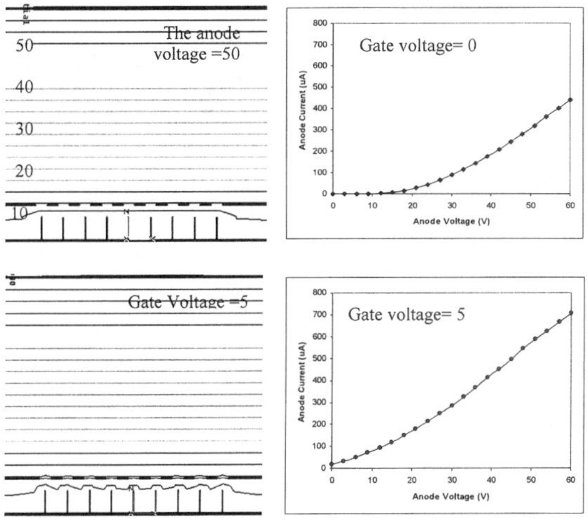

Figure 7: Left: The effect of gate voltage on the anode field penetration and field emission. Right: Experimental data for the effect of gate voltage on the emission current. All emission currents are highly controlled by the gate voltage.

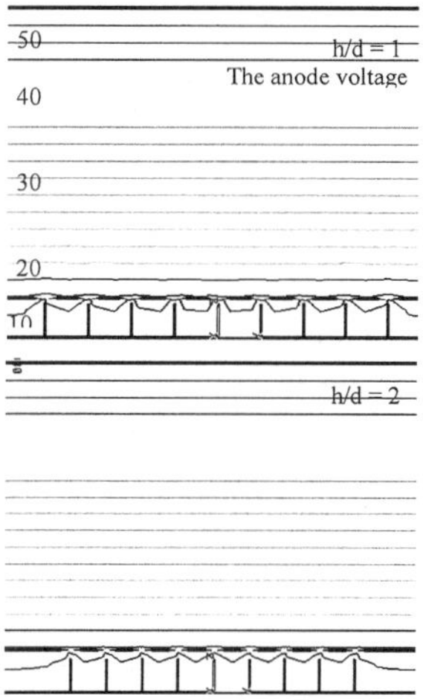

Figure 8: Simulation results for effect of gate plate on the screening effect and field enhancement factor in the structure of gated filed emitters. CNT's can be placed closer in this structure without a considerable screening effect.

Using the gate opening with a large diameter causes the high anode field penetrate near the CNT tip which in turn leads to an increase in the emission current. The preceding discussion has been based on a cluster of 3-by-3 array which has CNT's with the distance about 300nm and the length of 2μm. As was mentioned above the filed penetration is related to the diameter of the gate openings. The other parameter which affects the field emission is the space between tip and the gate openings. The results of Fig. 6 show the dependence of the field emission and the spacing between the tip and the gate opening. In other words, if the distance between the CNT tip and the gate opening increases, the penetration of the anode field through the opening is reduced and anode would lose its control on the emission current. This is an important conclusion to observe that the gate voltage control the emission current and the anode-cathode voltage has little effect on the emission current. An important result of Fig. 6 is that the gate influences the field and results in the reduction of the screening effect. In this simulation the gate voltage is set to be zero. The diameter of the gate opening and space between the CNT and gate are controlled in the fabrication process by using the mask with larger opening windows to pattern the gate and changing the gate oxide thickness. In our self-define process each CNT has its own gate surrounding it and the patterning of each gate is achieved with no need to a lithography process. Therefore the spacing between the gate and the tip of each CNT is essentially the thickness of the oxide layer which is well defined during the fabrication process.

Fig. 7 shows the effect of the gate voltage on the emission current, field penetration and enhancement factor. Also the results of simulation have been compared with the experimental data. The experimental data are obtained using a Keithley 2361 parameter analyzer by measuring the anode current with respect to the gate/cathode and anode/cathode voltages. Referred to the experimental data of Fig. 7 (right) at low anode voltages only the gate voltage causes an emission of electrons form the CNT tip. Under such conditions the emission is highly controlled by the gate voltage. The penetration of the electric field due to the anode electrode through the opening is limited, so a small current is observed. If the gate voltage increases the emission current shows a rapid rise with the anode voltage.

Finally we have investigated the field enhancement factor by changing the intertube distances. Figure 8 show the equipotential lines of the electrostatic field in the structure of CNT triode filed emitters. It is evident that the screening effect cannot occur in the intertube distance less than the CNT length resulting to decrease the intertube distance to the closet distance.

Conclusion

The operation of field emission devices strongly depends on the shape and arrangement of the CNT emitters. In intertube distances more than the length of CNT the operation or the emission current density increases because of the reduction of screening effect. Also the metal gate can reduce the screening effect and lead to a higher control of the gate voltage on the emission current. Based on the data obtained from the simulation results the best performance is obtained if the distance between nanotubes is set to be the same as the length of nanotubes. Regarding to the self defined process to realize the CNT tip from the bi-layers oxide-metal, each CNT has its own gate and its current can be controlled with its individual surrounded gate. As observed from the simulation results the gate plate around the CNT noticeably reduces the screening effect. Therefore, the metal gate would cause the nanotubes close together with distances less than the CNT length resulting in increasing the density of the nanotubes emitters and emission current density.

References:

[1] K. L. Jensen, R. H. Abrams, and R. K. Parker: J. Vac. Sci. Technol. B, Vol. 16-2 (1998), p. 749

[2] C. A. Spindt, C. E. Holland, P. R. Schwoebel, and I. Brodie: J. Vac. Sci. Technol. B, Vol. 16-2 (1998), p. 758

[3] A. Wisitsora-at, W.P. Kang, J.L. Davidson, Q. Li, J.F. Xu, and D.V. Kerns: The 11th International Vacuum Microelectronics Conference, North Carolina State University, Raleigh, NC, USA, (1998)

[4] K.C. Holmes, A. Wisitsora-at, T.G. Henderson, J.L. Davidson, W.P. Kang, and V. Pulugurta: the 196th Meeting of Electrochemical Society, Hawaii, USA., (1999)

[5] L. Frank and I. Mtillerova: Journal of Electron Microscopy, Vol. 48-3 (1999), p.205-219

[6] J. Park, J. D. Lera, H. J. Choi, G. H. Buh, C. J. Kang, J. H. Jung, S. S. Choi, D. Jeon, and Y. Kuk: J. Vac. Sci. Technol. B, Vol. 16-2 (1998), p. 826

[7] A. Fernandez, H. T. Nguyen, J. A. Britten, R. D. Boyd, M. D. Perry, D. R. Kania, and A. M. Hawryluk: J. Vac. Sci.Technol. B, Vol. 16-2 (1998), p. 729

[8] A. Naeemi, R. Sarvari, and J. D. Meindl: IEEE (2004), p. 8681

[9] L. Nilsson, O. Groening, C. Emmenegger, O. Kuettel, E. Schaller, and L. Schlapbach: Appl. Phys. Lett, Vol. 76 (2000), p. 2071

[10] J. Koohsorkhi, H. Hosseinzadegan, and S. Mohajerzadeh: Journal of Carbon, Vol. 44 (2006), p. 2797

[11] B. Arvan, A. Khakifirooz , R. Tarighat, S. Mohajerzadeh , A. Goodarzi, E. Asl Soleimani, and et al: Mater Sci Eng B, Vol. 109-1 (2003), p.17

[12] J. Koohsorkhi, S. Mohajerzadeh, Y. Abdi, and P. Hashemi: Proceeding of Mater. Res. Soc. Vol. 963 (2007)

[13] N. De Jonge, N.; J.-M. Bonard: Philosophical Transactions of the Royal Society London, Series A, Vol. 362 (2004), p. 2239

[14] J. M. Bonard, K. A. Dean, B. F. Coll, C. Klinke: Phys. Rev. Lett. Vol. 89 (2002), p. 197602

[15] C. Klinke: *Analysis of Catalytic Growth of Carbon Nanotubes* (EPFL, 2003)

[16] Y. Wei, C. Xie, K.A. Dean, B.F. Coll: Appl. Phys. Lett. Vol. 79 -27 (2001), p.4527

Solid State Phenomena Vol. 154 (2009) pp 89-94
online at http://www.scientific.net

FePt nanorods and nanowires for novel ferrofluids

Zhenyu Wu[1,a], Hailong Dong[1], Andrey Chuvilin[2], Ulf Wiedwald[3], Luyang Han[3] and Carl E. Krill III[1,b]

[1]Institute of Micro and Nanomaterials, Ulm University, Ulm, Germany

[2]Electron Microscopy Group of Materials Science, Ulm University, Ulm, Germany

[3]Institute of Solid State Physics, Ulm University, Ulm, Germany

[a]zhenyu.wu@uni-ulm.de, [b]carl.krill@uni-ulm.de

Keywords: FePt, nanorod, nanowire, ferrofluid, ordering transformation, superparamagnetism.

Abstract. The preparation of a novel ferrofluid out of one-dimensional magnetic structures is attempted by the reverse-micelle-based wet-chemical formation of FePt nanowires and nanorods. The latter were found to manifest lengths ranging from 25 to 250 nm, with an average diameter of only 1.5 to 2 nm. X-ray diffraction and high-resolution TEM revealed a polycrystalline microstructure, with individual grains having the disordered fcc structure in the as-prepared state; a thermally induced transition to the chemically ordered phase was observed around 550°C. The samples were determined to be superparamagnetic at room temperature, with the blocking temperature measured to be ~26 K.

Introduction

FePt *nanoparicles* have been studied extensively in recent years because of their high magnetoanisotropy and good chemical stability [1,2]. There are numerous promising applications of these materials in areas as diverse as ultra-high-density data storage [3–7], high-performance permanent magnets [8] and catalysis [9,10]. The one-dimensional morphology of FePt *nanorods* and *nanowires* represents a degree of freedom that may open up additional areas of application based on exploiting the shape anisotropy of the magnetic particles, such as in novel ferrofluids or magnetotransport devices [11]. For instance, it is believed that the phenomenon of shear thinning—which restricts the measurable magnetoviscosity of conventional ferrofluids to very low shear rates [12]—could be suppressed by replacing the constituent spherical nanoparticles with magnetic nanorods [11]. Such a "nanorod ferrofluid" might be useful as the working substance in an active damping system or in any other application relying on switchable force transfer via a liquid.

Of course, the preparation of a useful volume of ferrofluid out of one-dimensional magnetic structures is feasible only if very large numbers of the latter can be synthesized without sacrificing control over the uniformity of the nanorod/nanowire shape. A quite promising method in this regard is the approach pioneered by Wang *et al.* [13] for preparing nanowires (NWs) and nanorods (NRs) of FePt. (Following the usual convention, we will refer to one-dimensional nanostructures having a length of 100 nm or longer as NWs, and those below 100 nm as NRs.) From colloidal chemistry [14], we know that at different concentrations, surfactant molecules tend to form micelles having different geometries, ranging from spherical to cylindrical or even ordered hexagonally packed cylindrical structures. Wang *et al.* [13] proposed using a mixture of oleylamine (OAM) and octadecene (ODE), because for this surfactant combination the length of the resulting cylindrically shaped reverse micelles can be tailored to the desired size and shape simply by tuning the ratio of OAM to ODE. The resulting one-dimensional structures behave like compartments for further chemical reactions. By loading platinum (II) acetylacetonate [$Pt(acac)_2$] and iron (III) pentacarbonyl [$Fe(CO)_5$] precursors into micellular compartments, followed by thermo-decomposition at elevated temperature, several groups have succeeded in preparing NWs and NRs of FePt by this approach, with diameters of 2–3 nm and lengths tunable from more than 200 down to 20 nm [13,15,16]. In this work we report our efforts to adapt this approach to the preparation of a ferrofluid out of FePt NWs and/or NRs, placing special emphasis on obtaining large quantities of uniformly sized anisotropic mangetic particles.

Experimental

Our synthesis method closely followed that described in Ref. [13]. All reagents—iron (III) pentacarbonyl ([$Fe(CO)_5$], ≥ 97%, Fluka), oleylamine (≥ 70%, Fluka), platinum (II) acetylacetonate ([$Pt(acac)_2$], ≥ 97%, Alfa Aesar), octadecane (≥ 90%, Alfa Aesar) and hexane (≥ 99%, Sigma Aldrich)—were used without further purification, and all reactions were carried out under nitrogen gas in order to avoid oxidation. A typical synthesis procedure was as follows: OAM and ODE (2 ml each) were mixed in a three-necked conical flask, which was placed in an oil bath and bubbled with nitrogen for one hour to remove any oxygen in the solution. [$Pt(acac)_2$] (0.04 g) was added and the mixture heated to 65°C with continuous magnetic stirring until the [$Pt(acac)_2$] particles were dissolved to form a light yellow solution. The latter was then heated to 120°C in 5 min and kept at this temperature for 30 min. Subsequently, [$Fe(CO)_5$] (31 μl) was injected into the hot solution. The temperature was raised to 160°C within 15 min and kept there for an additional 30 min. After reaction, the solution was allowed to cool to room temperature naturally.

The reaction products were collected by washing away the remaining salts and surfactants. NWs and NRs were precipitated by adding ethanol (600 μl) into 200 μl of the as-synthesized solution and separated by centrifugation at 6000 rpm for 10 min. The yellow-brown supernatant was discarded. The black precipitate was then redispersed in hexane (300 μl) with the help of ultrasound for 3 min. The precipitation and redispersion steps were repeated three times in order to remove all impurities. Finally, the precipitate was dispersed in hexane (1000 μl) in the presence of oleic acid (OA, 0.5 μl) and OAM (0.5 μl), which stabilize the solution against agglomeration and sedimentation.

In addition to the results reported in Ref. [13], we found that this synthesis procedure generated spherical nanoparticles (NPs) as a reaction by-product. In order to separate NPs from the NWs and NRs, 100 μl of prepared solution was diluted by adding 899 μl hexane, 0.5 μl OA, 0.5 μl OAM and 500 μl ethanol. The diluted solution was centrifuged at 1000 rpm for 30 min. All of the supernatant was transferred to another container, and the precipitate was redispersed in 1000 μl hexane in the presence of 0.5 μl OA and 0.5 μl OAM. The precipitation and redispersion steps were repeated several times until the supernatant became colorless. Both the supernatant and the redispersed precipitate were retained for further analysis.

FePt NWs/NRs were characterized using a conventional transmission electron microscope (TEM, Philips EM 301, 80 kV) and a high-resolution TEM (HR-TEM, FEI Titan, 300 kV, C_S-corrected). The elemental composition of samples was measured in a scanning electron microscope (SEM, LEO 1550) by energy-dispersive x-ray spectroscopy (EDX). Phase analysis of powder samples was carried out by x-ray diffraction (XRD, PANaltyical X'Pert Pro, Co-K_α radiation, λ=0.179 nm). The thermally induced ordering transformation of the as-prepared FePt phase was observed *in situ* by XRD of a sample mounted inside a vacuum oven (mri Physikalische Geräte GmbH) at a base pressure of 10^{-6} mbar. Magnetic measurements were carried out using a SQUID magnetometer (Quantum Design).

Results and Discussion

Morphology. A typical TEM micrograph of FePt NWs and NRs after washing is shown in Fig. 1(a). In addition to the expected presence of NWs/NRs, there are also a considerable number of spherical NPs in the sample. A possible explanation for this finding would be the inability of some FePt nuclei to grow beyond the thermodynamically preferable spherical shape, owing to local defects or fluctuations in the shape of the soft micellular templates. Fig. 1(b) shows a TEM micrograph of the supernatant after the separation procedure described in the previous section, whereas Fig. 1(c) illustrates the final precipitate. By choosing proper solvents and optimizing the

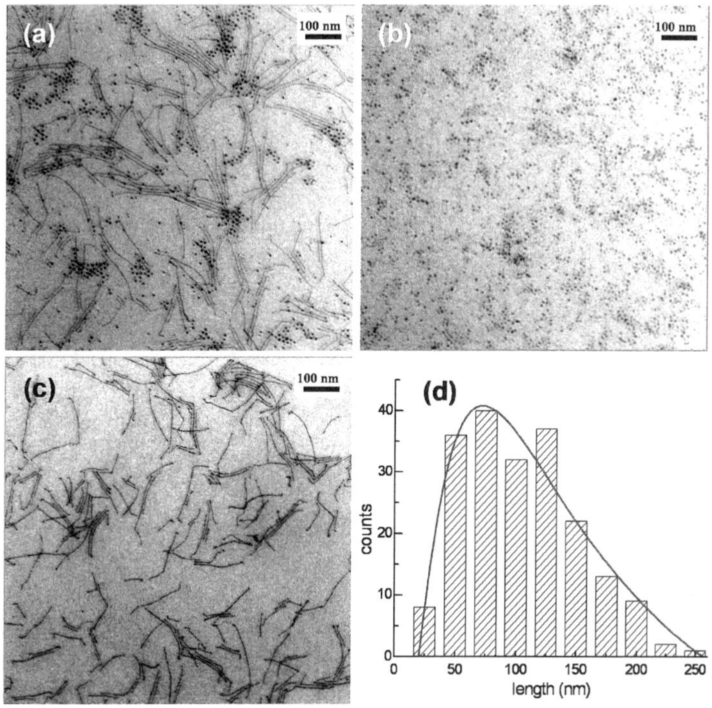

Fig. 1. TEM micrographs of FePt NWs and NRs: (a) directly after washing, (b) spherical NPs after separation, and (c) NWs/NRs after separation. (d) Length distribution of the NWs/NRs in (c); the solid curve is a least-squares fit of a lognormal distribution function to the histogram.

parameters of the separation procedure, we were able to remove the spherical NPs completely, leaving solely NWs and NRs. The length distribution of the NWs and NRs shown in Fig. 1(c) is plotted in Fig. 1(d), from which we conclude that most of the wires are between 50 and 150 nm in length.

Microstructure. The overall composition of the NWs/NRs was $Fe_{55}Pt_{45}$, as determined by EDX spectroscopy performed on powder samples. XRD of as-synthesized NWs/NRs revealed the presence of the chemically disordered face-centered cubic (fcc) phase of FePt at room temperature (RT) and also following a 350°C anneal for 1 h (Fig. 2). The broad peaks evident in the RT scan originate from the presence of nanometer-sized coherent scattering regions, which are a direct consequence of the small diameter of the NWs/NRs. After correcting for the instrumental contribution to the measured full width at half maximum (FWHM) of the (111) Bragg peak, we use the Scherrer formula [17] to calculate an average crystallite size of 3.5 nm. Thermal annealing in vacuum at and above 550°C induces a transformation to the chemically ordered face-centered tetragonal (fct) structure, as evidenced by the peak splitting and new peaks emerging in the XRD patterns at 550°C and 650°C. The peaks become sharper at higher temperatures, which indicates the occurrence of grain growth of the FePt crystallites and, in all likelihood, their sintering into larger aggregates. From the FWHM of the (111) peak in the 650°C pattern, we estimate an average grain size of about 160 nm using the Scherrer formula.

High-resolution TEM images (Fig. 3, left) show good crystallinity of the FePt NWs/NRs. The selected area diffraction (SAD) image confirms that the NWs/NRs are polycrystalline with an fcc lattice. On the outside surface of each wire, we see evidence for the presence of an amorphous layer having a thickness of a few angstroms. This may arise from oxidation of Fe atoms after exposure to air during TEM sample preparation. In the right-hand image of Fig. 3, the atomic planes of adjacent

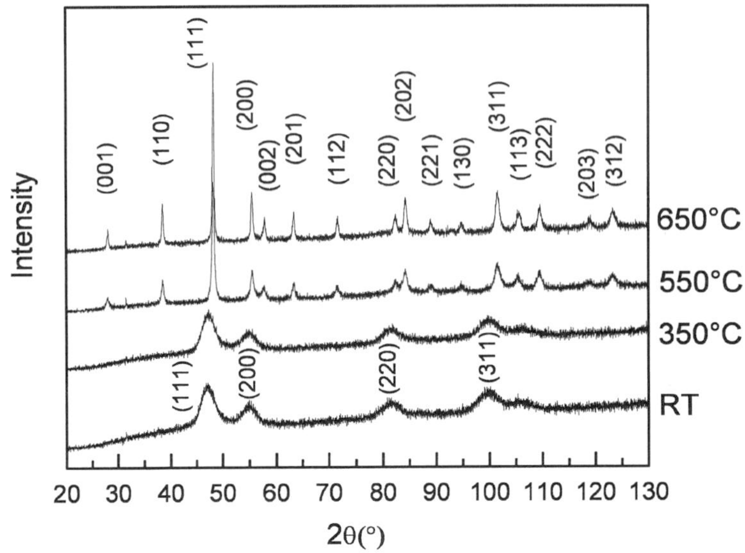

Fig. 2. X-ray diffraction scans of FePt NWs/NRs, annealed under vacuum for 1 h at the indicated temperatures.

crystallites with different orientations within a single NW are resolved. In region (a) the measured spacing of 0.221 nm is nearly identical to the lattice spacing of 0.223 nm that was calculated from the (111) peak position in the XRD pattern of Fig. 2. In region (b) of Fig. 3, not only are lattice planes resolved, but also atomic columns; again, the contrast is consistent with that of (111) planes, as the angle between two sets of equivalent lattice planes is 70.5°—the interplanar angle between (111) planes in an fcc lattice. In region (c), the interplanar spacing of 0.133 nm corresponds to that expected for (220) planes.

Based on a tendency evident in the HR-TEM images for one set of (111) planes to align parallel to the long axis of a NW, we can estimate the average size of a coherent scattering region for the (111) Bragg reflection. In an fcc lattice, there are four sets of distinct (111) planes, all oriented at an angle of 70.5° relative to one another. By simple geometry, we calculate the average distance L from one edge of a stack of parallel (111) planes to the other when they are located within a NW of diameter ϕ to be

$$L = \frac{\phi}{4}\left[1 + \frac{3}{\cos(70.5°)}\right]. \tag{1}$$

Inserting $\phi \approx 1.5$ nm into Eq. 1, we obtain $L \approx 3.7$ nm, which agrees quite well with the value for the same quantity (3.5 nm) that was estimated from the FWHM of the (111) XRD peak in the RT scan of Fig. 2 inserted into the Scherrer formula (see above).

Magnetic properties. As can be seen in the magnetometry measurements of Fig. 4, the as-synthesized FePt NWs/NRs are superparamagnetic at room temperature, while at 10 K their hysteresis loop manifests a coercivity of about 500 Oe (50 mT)—indicative of ferromagnetism—and a much higher saturation magnetization. The temperature-dependent magnetic moment was recorded during zero-field-cooling (ZFC) and field-cooling (FC) scans between 10 and 300 K in a 100 Oe applied magnetic field. Comparison of the ZFC and FC curves reveals a blocking temperature for the superparamagnetic-ferromagnetic transition of ~26 K.

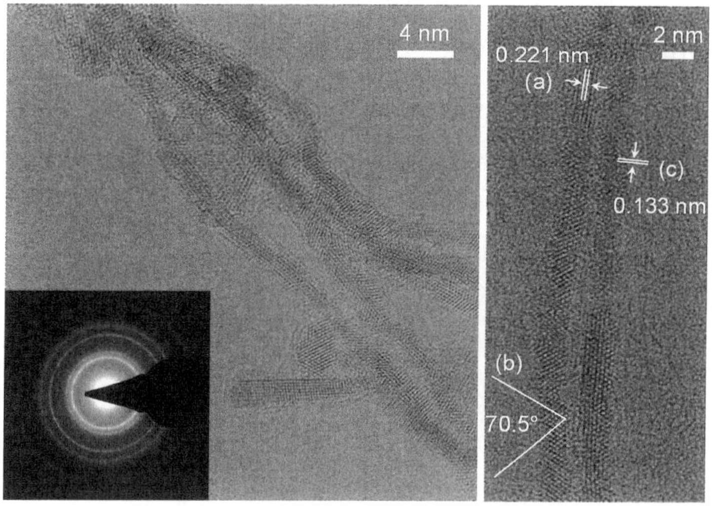

Fig. 3. (Left) HR-TEM image of FePt NWs; inset: selected area diffraction (SAD) of an agglomerate of NWs. (Right) Atomic planes of polycrystalline NWs, imaged by HR-TEM. See text for details.

According to our experience studying the rheological properties of conventional ferrofluids, a sample magnetization of about 5×10^{-4} emu/μl leads to a *magnetoviscosity* signal—i.e. the viscosity increase induced by the application of a stationary magnetic field [18]—of around 20% (relative viscosity change) for a carrier fluid consisting of a water/diethylene glycol solution. A suspension of FePt NWs/NRs at a volume fraction of 0.45% was measured to have a room-temperature magnetization of around 4.8×10^{-4} emu/μl at an applied magnetic field of 1 T, suggesting that it ought to be possible to study the influence of particle shape anisotropy on magnetoviscosity by subjecting FePt NW/NR suspensions to magnetic fields of comparable strength. In future experiments, we will test this prediction and also attempt to increase the total sample magnetization still further by raising the volume fraction of FePt NWs and NRs in suspension; this latter step should drastically lower the field strength for the onset of significant magnetoviscous effects, which would be highly desirable from the standpoint of technological application.

Conclusions

The reverse-micelle approach to preparing one-dimensional FePt structures [13] was found to be suitable to the large-scale production of magnetic nanorods (NRs) and nanowires (NWs) for application as constituents of novel ferrofluids made out of anisotropic magnetic particles. Morphological characterization revealed a rather broad distribution of particle lengths, ranging from 25 to 250 nm, with the majority lying between 50 and 150 nm; however, the NR/NW diameter of 1.5–2 nm was independent of the length. According to x-ray diffraction and high-resolution TEM studies, the FePt NRs and NWs manifested a polycrystalline microstructure consisting of grains of the chemically disordered fcc phase. Thermal annealing at 550°C induced a transformation to the ordered fct phase, which was accompanied by grain growth and sintering. At temperatures above ~ 26 K, the NRs and NWs were found to be superparamagnetic, but even magnetic fields as strong as 3 T were not able to saturate the sample. Extrapolating from the room-temperature hysteresis curve, we conclude that a ferrofluid made out of FePt NRs and/or NWs would show a measurable magnetoviscosity only at applied magnetic fields in excess of ~1 T, unless a NR/NW volume fraction on the order of 2% or higher could be achieved in a stable suspension.

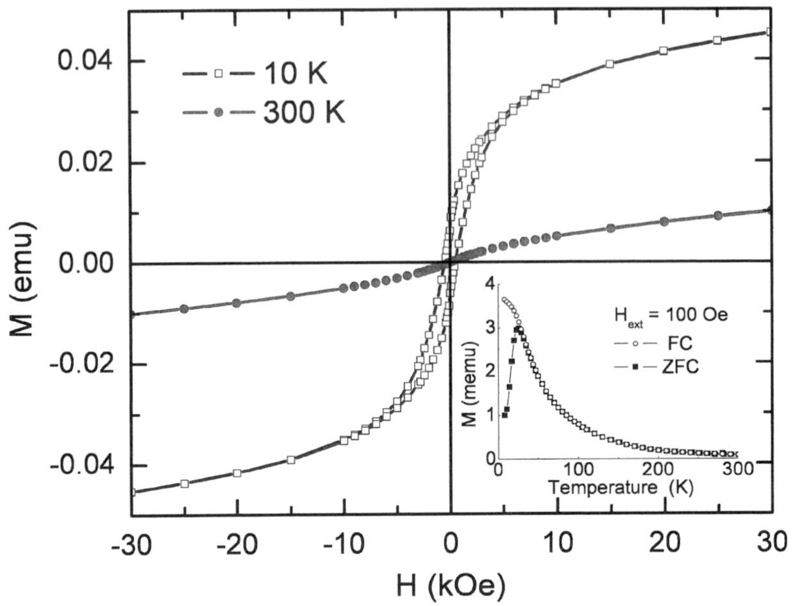

Fig. 4. Magnetic hysteresis loops of a powder sample of as-synthesized FePt NWs/NRs measured at 10 and 300 K; inset: zero-field-cooling (ZFC) and field-cooling (FC) measurements performed on the same sample, indicating a value of the blocking temperature of

Acknowledgments

We gratefully acknowledge fruitful discussions with Dr. Ulrich Ziener of the Institute of Organic Chemistry III at Ulm University. Financial support for this work was provided by the Deutsche Forschungsgemeinschaft within the context of Research Priority Program 1165 "Nanowires and Nanotubes".

References

[1] S. Sun, C. B. Murray, D. Weller, L. Folks, A. Moser: Science, Vol. 287 (2000), p. 1989.
[2] S. Sun: Adv. Mater., Vol. 18 (2006), p. 393.
[3] V. Nandwana, K. E. Elkins, N. Poudyal, G. S. Chaubey, K. Yano, J. P. Liu: J. Phys. Chem. C, Vol. 111 (2007), p. 4185.
[4] C. B. Rong, D. R. Li, V. Nandwana *et al.*: Adv. Mater., Vol. 18 (2006), p. 2984.
[5] C. B. Rong, V. Nandwana, N. Poudyal *et al.*: J. Appl. Phys., Vol. 102 (2007), p. 023908.
[6] M. Chen, J. Kim, J. P. Liu, H. Y. Fan, S. H. Sun: J. Amer. Chem. Soc., Vol. 128 (2006), p. 7132.
[7] A. Ethirajan, U. Wiedwald, H. G. Boyen *et al.*: Adv. Mater., Vol. 19 (2007), p. 406.
[8] G. Srajer, L. H. Lewis, S. D. Bader *et al.*: J. Mag. Mag. Mater., Vol. 307 (2006), p. 1.
[9] J. Jiang, A. Kucernak: J. Electroanal. Chem., Vol. 520 (2002), p. 64.
[10] T. Toda, H. Igarashi, H. Uchida, M. Watanabe: J. Electrochem. Soc., Vol. 146 (1999), p. 3750.
[11] R. Birringer, H. Wolf, C. Lang, A. Tschöpe, A. Michels: Z. Phys. Chem., Vol. 222 (2008), p. 229.
[12] S. Odenbach: J. Phys. Conden. Matter, Vol. 16 (2004), p. R1135.
[13] C. Wang, Y. Hou, J. Kim, S. Sun: Angew. Chem. Int. Ed., Vol. 46 (2007), p. 6333.
[14] S. Forster, M. Antonietti: Adv. Mater., Vol. 10 (1998), p. 195.
[15] M. Chen, T. Pica, Y. B. Jiang *et al.*: J. Amer. Chem. Soc., Vol. 129 (2007), p. 6348.
[16] P. Narayan, S. C. Girija, N. Vikas, R. Chuan-bing, Y. Kazuaki, J. P. Liu: Nanotechnology, Vol. 19 (2008), p. 355601.
[17] B. D. Cullity: *Elements of X-Ray Diffraction*. (Addison-Wesley Publishing Company, Inc., 1978), p. 102.
[18] S. Odenbach: *Magnetoviscous Effects in Ferrofluids*. (Springer, 2002).

Solid State Phenomena Vol. 154 (2009) pp 95-100
online at http://www.scientific.net

Formation of Metal Silicide Nanodots on Ultrathin SiO_2 for Floating Gate Application

S. Miyazaki, M. Ikeda, K. Makihara, K. Shimanoe and R. Matsumoto
Graduate School of Advanced Sciences of Matter, Hiroshima University
1-3-1 Kagamiyama, Higashi-Hiroshima 739-8530, Japan
semiya@hiroshima-u.ac.jp

Keywords: Silicide, Nanodots, Floating Gate, Charge Storage, H_2 plasma

Abstract. We demonstrated a new fabrication method of Pt- and Ni-silicide nanodots with an areal density of the order of ~10^{11} cm^{-2} on SiO_2 through the process steps of ultrathin metal film deposition on pre-grown Si-QDs and subsequent remote H_2 plasma treatments at room temperature. Verification of electrical separation among silicide nanodots was made by measuring surface potential changes due to electron injection and extraction using an AFM/Kelvin probe technique. Photoemission measurements confirm a deeper potential well of silicide nanodots than Si-QDs and a resultant superior charge retention was also verified by surface potential measurements after charging to and discharging. Also, the advantage in many electron storage per silicide nanodot was demonstrated in C-V characteristics of MIS capacitors with silicide nanodots FGs.

Introduction

Charge storage in metallic nanodots embedded into a dielectric layer has received much attention because of their potential advantages in storage capacity in comparison to semiconducting quantum dots[1-3]. In particular, metallic nanodots with a high work function relative to the electron affinity of the substrate serve a deep potential well for electrons, which enables us to decrease the tunnel oxide thickness and results in faster writing and erasing operation, while maintaing nonvolatility. In the floating gate application of metallic nanodots, one of the major technological issues is to form uniformly metallic nanodots with an areal density higher than $10^{12}cm^{-2}$ without degrading gate oxide reliability caused by metal diffusion [4]. We have developed a new fabrication method of nanometer-scale metal dots on thermally-grown ultrathin SiO_2 with a fairly uniform size distribution and a high areal density, in which ultrathin films of oxidation-resistant metals such as Ni, Pt and Pd are exposed to remote H_2 plasma at room temperature [5-7].

In this work, we extended our research to form Ni- and Pt-silicide dots showing good thermal stability in comparison to elemental metal dots. Chemical and electronic structures of metalic dots prepred in this work are reported and, in metal-insulator-semiconductor (MIS) capacitors with dots as a floating gate, electrical charaging characteristics of dots are also examined to demonstate their potential in memory applications.

Sample Preparation

After conventional wet-chemical cleaning steps of p-Si(100) wafers, ~3.6nm thick SiO_2 layers were grown at 1000C in dry O_2, and hemispherical Si quantum dots (QDs) were self-assembled on SiO_2/Si(100) at 560°C by controlling the early stages of CVD using pure SiH_4[8]. Ultrathin Pt and Ni films in the thickness range of 1.4~4.0 nm were formed on Si-QDs/SiO_2/Si(100) and on SiO_2/Si(100) as reference at room temperature by Ar sputtering and thermal evaporation, respectively. Subsequently, the samples were exposed to remote plasma of pure H_2 without any extra heating. The plasma was generated in a quartz tube with a diameter of 10cm by inductively-coupling with an external single-turn antenna connected to a 60MHz generator [5]. In typical conditions of remote H_2 plasma treatment, VHF power, gas pressure and exposure time were set at 400W, 0.1Torr and 5min, respectively. The substrate was placed on a quartz susceptor at a distance of 23cm away from the position of the antenna to reduce efficiently ion incidence from the plasma to the sample surface.

In the fabrication of MIS capacitors, 15~20nm-thick SiO_2 layers were prepared as a control oxide at 350°C on H_2-plasma treated samples by remote plasma-enhanced CVD in which O_2 gas diluted with Ar was excited separately with an inductively-coupled plasma and reacted to SiH_4 [6], and then Al gate with a diameter of 1mm was fabricated at room temperature by thermal evaporation through a stencil mask.

Results and Discussion

We firstly evaluated changes in the surface morphology for each process step by atomic force microscopy (AFM) as shown in Fig. 1. AFM images confirm the formation of Si-QDs with an areal dot density of $4x10^{11}cm^{-3}$ on the SiO_2 surface and almost no change in the surface morphologies with conformal deposition of a ~2nm-thick Pt layer. Also, by remote H_2-plasma treatment, the dot height is slightly increased but the dot densities almost remain unchanged. Considering the fact that ~3.8nm-thick Pt layer on SiO_2 is changed into Pt nanodots with a number density of $2.4x10^{11}$ cm-3 by the remote H_2-plasma treatment with the same condition as show in Fig. 2, it is likely that the Si-QDs act as a sink for agglomeration. Since the Pt dot formation on SiO_2 by remote H_2-plasma treatment takes place even when a grounded stainless mesh grid of 1mm spacing is placed 1cm above the sample during the remote H_2-plasma treatment as schematically illustrated in Fig. 2 (d), the incidence of hydrogen radicals to the Pt surface and their surface recombination to create energetically stable H_2 molecules is thought to be responsible for the enhanced surface migration of Pt atoms and the resultant agglomeration to minimize the surface free energy. Very similar results were obtained for

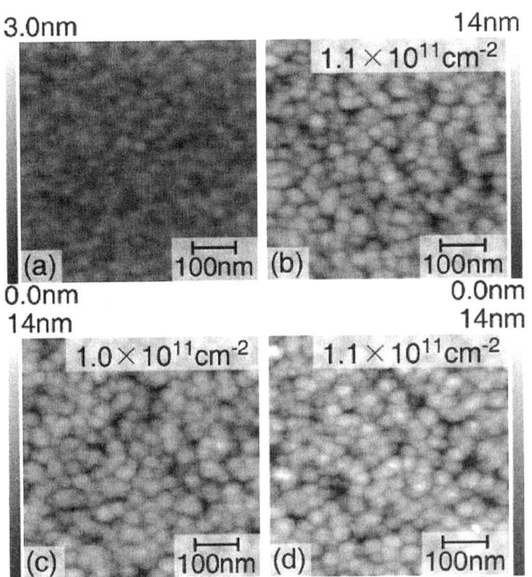

Fig. 1 AFM images of as-grown SiO_2 (a), Si-QDs (b), ~2nm-thick Pt film deposited on Si-QDs before (c) and after remote H_2-plasma treatment at a VHF power of 400 W(d). H_2 pressure and substrate temperature were maintained at 0.1 Torr and RT, respectively.

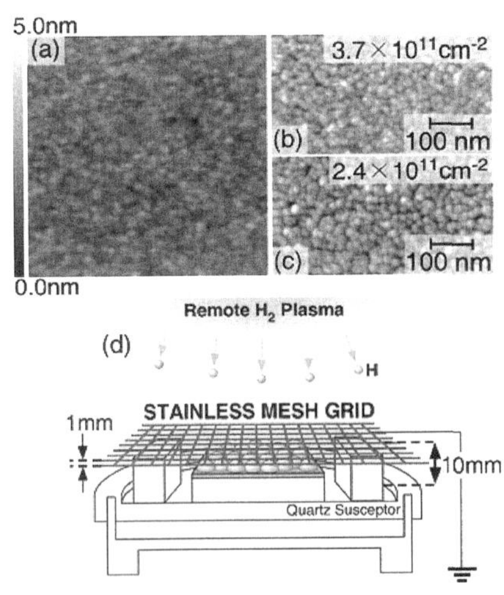

Fig. 2 AFM images of as-deposited ~3.8nm-thick Pt film on SiO_2 (a) and after remote H_2 plasma treatment with (b) and without (c) stainless mesh grid. H_2 pressure, VHF power and substrate temperature were maintained at 0.1Torr, 400W and RT, respectively. A schematic view of the configuration of mesh grid during remote H_2-plasma treatment (d).

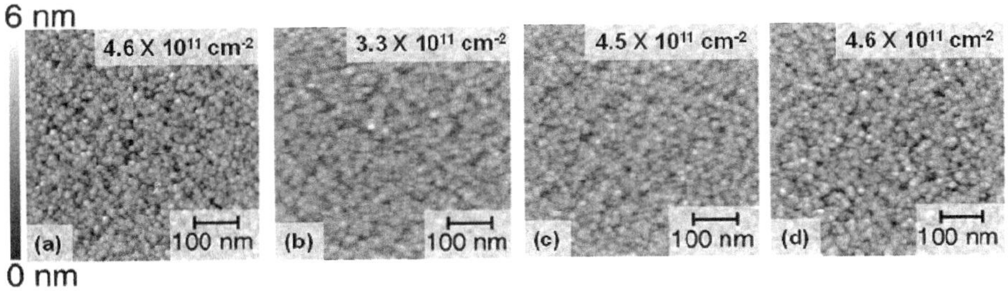

Fig. 3 AFM images of as-prepared Si QDs on SiO_2/Si(100) (a), after ~1.4nm-thick Ni film deposition (b), subsequent remote H_2 plasma treatment (c) and then complete removal of un-reacted Ni from H_2 plasma treated surface by dipping in a HCl solution.

Ni case. As shown in Fig. 3, no significant changes in surface morphology and dot densities were observed through the process steps.

Next, chemical and electronic structures of the samples after H_2 plasma treatment were examined by high-resolution X-ray photoemission spectroscopy (XPS) using monochromatized AlKα (1486.6eV) radiation. Figures 4 and 5 show valence band spectra taken after exposuring Pt/ and Ni/Si-QDs on thermally-grown SiO_2 to remote H_2 plasma, respectively. In both cases, distinct differences in valence band spectrum between sample and pure metal film are clearly present especially in the region near the valence band top. In the case of Pt, shown in Fig. 4, because of the very thin SiO_2 layer (down to ~3.4nm), photoelectrons from the bottom Si substrate contribute to the measured valence band spectra. To extract the signals from dots, the signals originating from the Si substrate and the SiO_2 layer were subtracted from the measured spectrum in consideration of the signal intensity and the binding energy of Si2p core line (Fig. 4 (b)). Considering the band bending in the underlying SiO_2/Si(100), which corresponds to a shift by 0.8eV toward the higher bind energy side, residual signals shown with open circles are likely to match with the valence band spectrum measured for Pt-silicide. Thus, we confirmed the silicidation of Si-QDs. In the Ni case shown in Fig. 5, signals originating from the underlying SiO_2 layer being sufficiently thicker than the photoelectron escape depth are only overlapped in the region deeper than ~5eV in binding energy. After removal of Ni oxide from the sample surface by dipping in a HCl solution, signals corresponding to Ni-Si

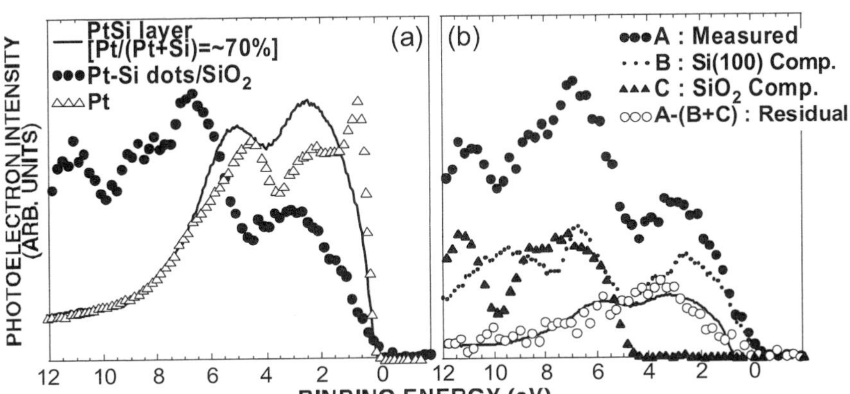

Fig. 4 Valence band (VB) spectra taken after remote H_2-plasma treatment of Pt-Si-QDs/~3.4nm-thick SiO_2/Si(100) (a), where VB spectra for pure Pt and Pt-silicide films are also shown as refereces, and the spectral deconvolution of measured VB spectrum **A** for the remote H_2-plasma treated sample (b). The Pt content in the Pt-silicide film is ~70% in Pt/(Pt+Si) ratio. In (b), **B** and **C** are deconvoluted components due to SiO_2 and Si(100) substrate, respectively, and residual signals (open circles) being attributable to the VB of H_2-plasma treated dots were compared with the shape of the VB spetrum of PtSi (solid line).

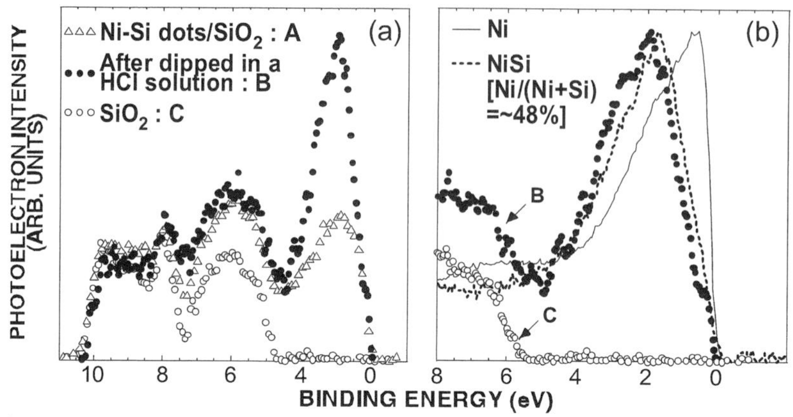

Fig. 5 Valence band (VB) spectra taken after remote H_2-plasma treatment of Ni-Si-QDs/~20nm-thick SiO_2 **A** and subsequent HCl etching **B** (a), where the VB spectrum of thermally-gowtn SiO_2 **C** is also shown as a reference, and comparison between the VB spectra taken for the H_2-plasma treated sample, Pure Ni and Ni-silicide films (b). The Ni content in the NiSi film is ~48% in Ni/(Ni+Si) ratio. In (a), the spectral intensities were normalized by Si2p intensities originating from SiO_2.

bonding states are clearly detected in the region within ~4eV in binding energy (Fig. 5 (a)). As seen in Fig. 5 (b), from the spectral shape near the top of the valence band, the formation of Ni-silicide nanodots with a composition close to monosilicide was verified. These are also supported by the analysis of the chemical shift measured in core lines and the estimation of chemical composition from intensity ratio of core lines. In addition, from the cut-off energy of photoemission near the lower limit in kinetic energy, the work functions of Pt- and Ni-silicide nanodots so prepared were determined to be ~5.1 and ~4.5eV, respectively, being fairly close to the values for Pt- and Ni-silicide films. All the photoemission data indicate that the incidence of hydrogen radicals promotes silicidation of Si-QDs even at room temperature.

To check electrical separation among silicide nanodots so-prepared, electron injection to and extraction from the nanodots were carried out by scanning the sample surface with an electrically-biased AFM probe tip in tapping mode at room temperature in clean room air and then the surface potential changes were systematically measured by means of a non-contact Kelvine probe technique [9]. In this experiment, a Rh-coated Si_3N_4 cantilever was used. For the as-prepared sample surface, a uniform surface potential image is obtained as seen in Fig. 6 (a). When the surface was scanned with the AFM tip positively biased with respect to the substrate in tapping mode, a distinct increase in the surface potential due to the electron extraction from the dots only in the scanned area was clearly observed. Subsequently, by scanning with the negatively biased AFM tip in the central part of the pre-charged region, the surface potential was decreased by electron injection.

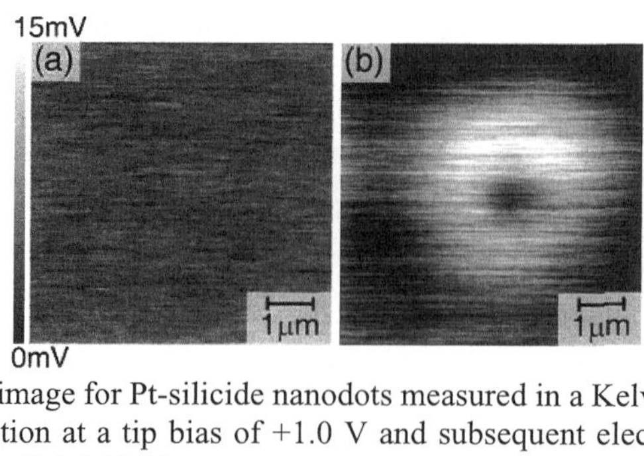

Fig. 6 Surface potential image for Pt-silicide nanodots measured in a Kelvin probe mode before (a) and after electron extraction at a tip bias of +1.0 V and subsequent electron injection to the pre-charged area at a tip bias of -1.0 V (b).

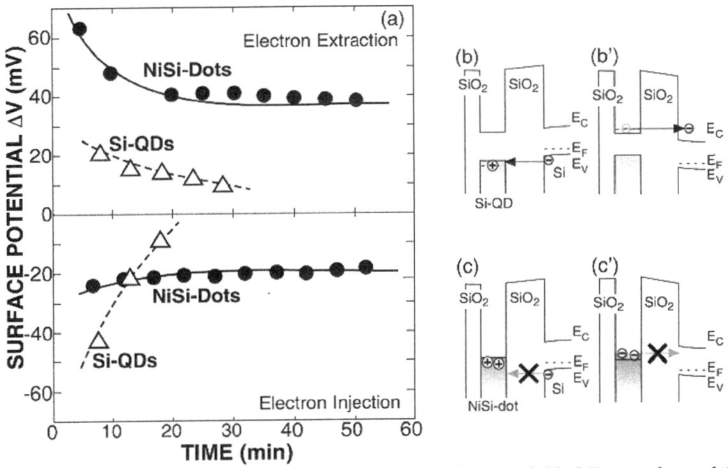

Fig. 7 Charge retention characteristics of Ni- silicide nanodots and Si-QDs evaluated from changes in the surface potential after electron extraction from and injection to the dots at tip biases of +3.0 and -3.0V, respectively (a). Energy band diagrams of negatively and positively charged Si-QDs (b, b') and NiSi-dots (c, c') on SiO_2/p-Si(100).

As a result, the potential image with a clear contract between charged and uncharged area was obtained as shown in Fig. 6 (b). In contrast, on the sample surface before remote H_2-plasma treatment, such a potential change was never detected, rather the potential image was always very uniform because of good electrical conduction on the sample surface. The result of Fig. 6 (b) indicates that the Pt-silicide nanodots are electrically isolated one to another. Charge retention in both positively and negatively charged states was evaluated as shown in Fig. 7 (a) in comparison to the results obtained for pure Si-QDs with almost the same size. Obviously, Ni-silicide nanodots show superior charge retention characteristic to Si-QDs, which is attributable to a deeper potential well reflecting the work function of Ni-silicide as seen in the energy band diagrams of Fig. 7 (b) and (c) .

Electrical charging to and discharging from silicide nanodots were also evaluated in MIS capacitors with silicide nanodots as a floating gate (FG). Figures 8 and 9 show capacitance-voltage (C-V) characteristics of the MIS capacitors with a Pt-silicide nanodots FG and a Ni-silicide ninodots

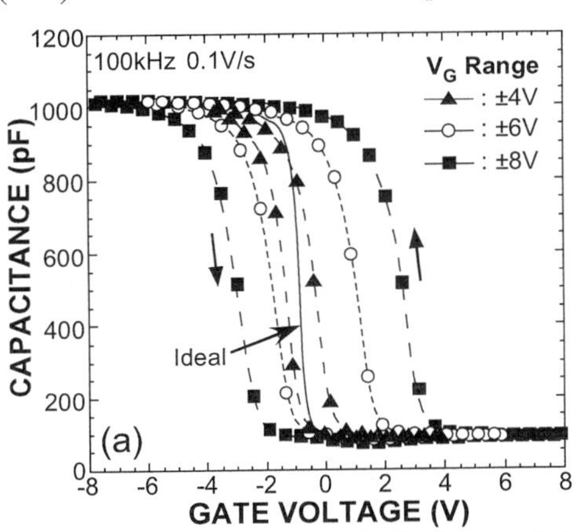

Fig. 8 Capacitance-voltage (C-V) characteristics of a MIS capacitor with a Pt-silicide nanodots floating gate measured at 100kHz with a gate voltage sweep rate of 0.1V/s. Ideal C-V curve is also shown as a reference.

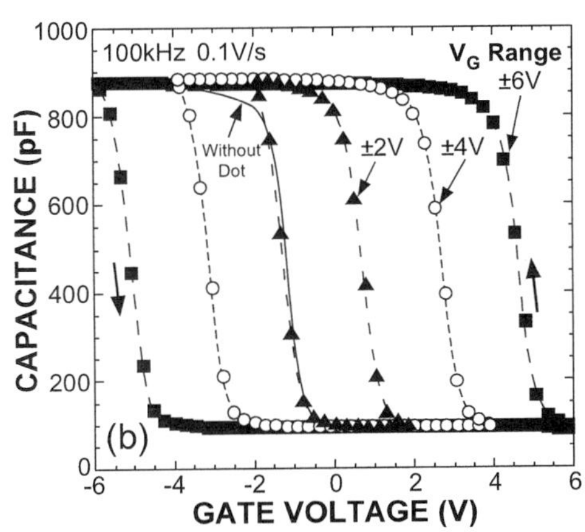

Fig. 9 C-V characteristics of MIS capacitors with and without a Ni-silicide nanodots floating gate measured at 100kHz with a gate voltage sweep rate of 0.1V/s.

FG, respectively, measured at 100kHz with a gate voltage (Vg) sweep rate of 0.1V/s, in which the gate voltage swing range was changed up to ±8V at room temperature. The C-V curve obtined for a MIS capacitor without FG, which was fabricated by the same process except the silicide nanodot foramtion, was also measured as a reference (Fig. 9) and no hystersis characteristic was observable. In contrast, for the silicide nanodts FG MIS capacitors, the flat-band voltage shift (ΔV_{FB}) was increased with larger gate voltages in each polarity. The positive ΔV_{FB} is caused by electron injection to the dots FG through the bottom tunnel oxide, in which the dots FG is negatively charged. On the other hand, the negative ΔV_{FB} is attributed to electron emission from the silicide nanodots, namely, positively charged dots FG. The differences in the amonts of ΔV_{FB} between the Pt-silicides and Ni-silicide nanodots FG are attributed to the differences in the dot density and the control oxide thickness. For the Si-QDs FG MIS capacitor which was fabricated skipping metal film deposition and other process steps were the same as thoes for the Ni-silicide nanodots FG MIS, the total ΔV_{FB} of ~ 1.2V corresponding to one electron storage per dot [10] was observed even with a Vg swing range of ±6V. Thus, the total ΔV_{FB} as high as 10V observed in Ni-silicide nanodots FG corresponds to the change by about 8 electrons per dot, which can be interpreted in terms of a deep potential well in Ni-silicide nanodots FG.

Summary

A novel fabrication method of Pt- and Ni-silicide nanodots with an areal density of the order of $\sim 10^{11}$ cm^{-2} has been demonstrated by the process steps of ultrathin metal film deposition on pre-grown Si-QDs and subsequent remote H_2 plasma treatments at room temperature. Electrical separation among silicide nanodots was verified from surface potential changes caused by electron injection and extraction utilizing an AFM/Kelvin probe technique. A deeper potential well of silicide nanodots than Si-QDs was confirmed by XPS measurements and superior charge retention was also shown by surface potential measurements as a function of time after charging and discharging. Many electron storage capability per silicide nanodot was also confirmed from C-V characteristics of MIS capacitors with silicide nanodots FGs.

Acknowledgment

This work was supported in part by Grant-in Aids for Scientific Research (A) No. 18206035 and for Scientific Research on Priority Area (458) No. 18063017 of MEXT, Japan.

References

[1] M. Takata, S. Kondoh, T. Sakaguchi, H. Choi, J-C. Shim, H. Kurino and M. Koyanagi, Tech. Digest of Int. Electron Devise Meeting (2003) p. 2251.
[2] C. H. Lee, J. Meteer, J. Narayana and E. C. Kan, J. of Electronic Materials 34 (2005) 1.
[3] D.B. Farmer and R.G. Gordona, J. of Appl. Phys. 101 (2007) 124503.
[4] J. T. Mayer, R. F. Lin and E. Garfunkel, Surf. Sci. 265 (1992) 102.
[5] K. Makihara, K. Shimanoe, M. Ikeda, S. Higashi and S. Miyazaki, Jpn. J. Appl. Phys. 47 (2008) 3099.
[6] M. Ikeda, K. Shimanoe, R. Matumoto, K. Makihara, S. Miyazaki, Proc. of 6th Int. Meeting for Future Electron Devices, Kansai (2008) p. 43.
[7] K. Shimanoe, K. Makihara, M. Ikeda and S. Miyazaki, Proc. of Asia-Pacific Workshop on Fundamentals and Applications of Advanced Semiconductor Devices (2008) p.77.
[8] S. Miyazaki, Y. Hamamoto, E. Yoshida, M. Ikeda and M. Hirose, Thin Solid Films 369 (2000) 55.
[9] K. Takeuchi, H. Murakami and S. Miyazaki, Proc. of ECS Int. Semicond. Tech. Conf. (2003) p. 1.
[10] A. Kohno, H. Murakami, M. Ikeda, S. Miyazaki and M. Hirose, Jpn. J. Appl. Phys. 40 (2001) L721.

Solid State Phenomena Vol. 154 (2009) pp 101-106
online at http://www.scientific.net

Nano-Silicon Sol-Gel Film Refraction Index Modulation with Femtosecond Laser

Antonela Dima[1,a], Massimo Gagliardi[2,b], Dun Liu[1,c], Walter Perrie[1,d], Craig J. Williams[1,e], Ivo Rendina[2,f], Geoff Dearden[1,g], Ken G. Watkins[1,h]

1. The University of Liverpool (UOL), Brownlow Hill, Liverpool L69 3GH, United Kingdom
2. Istituto per la Microelettronica e Microsistemi (IMM-CNR), via P. Castellino 111, Napoli 80131, Italy

[a] antonela.dima@liv.ac.uk, [b] Massimo.gagliardi@na.imm.cnr.it, [c] dun.liu@liv.ac.uk,[e] c.j.williams@liv.ac.uk,[f] ivo.rendina@na.imm.cnr.it, [g] g.dearden@liv.ac.uk, [h]k.watkins@liv.ac.uk

Keywords: silicon nanoparticles, sol-gel, waveguides, laser direct writing

Abstract: Patterned structures were created by exposing SiO_2 sol-gel films containing nano-silicon particles to a Clark MXR CPA-2010 fs laser (387 nm). A refractive index variation of 0.2 was obtained, similar to that of polymer films, however in an entirely superior stability class (structural, chemical, thermal, radiation, etc). The useful optical range of refractive index modulation is beyond 800 nm, respectively near-IR. Material characteristics were investigated with atomic force microscopy (AFM), Raman spectroscopy and spectro-ellipsometry measurements. Material properties were also investigated on different substrates in order to determine the influence of substrate type in laser processing.

Introduction

Direct laser writing techniques gained popularity in the last decade for a variety of applications, ranging from planar optical waveguides (needing no complex photolithographic processing) to material modulation. One of the techniques used is based on the focusing UV irradiation [1,2] onto Germania doped planar films deposited on silica – in order to gain a refractive index increase in the region exposed and form a "laser-written" waveguide. The refractive index modification forms the core of the waveguide, achieved refractive index changes being on the order of 10^{-3}. Femto-second lasers [3–5] also cause change in the structure of the silica – in turn itself contributing to a refractive index increase, of the same order ($\sim10^{-3}$). The modulation depends on exposure time, optical power and beam focusing. Erbium doped silica is a prime candidate for such techniques given its commercial availability and better properties for laser writing, permitting a relatively high modulation [6].

Separately from these investigations, the study of materials and devices containing nano-silicon crystals brought new devices with superior properties – such as intense light emission, making them potent candidates for Si-based opto-electronics. Such technological options are: nanoscale porous silicon (NPS), Si nanocrystalline embedded SiO_2 (SiO_x, $x < 2.0$) matrices (nc-Si/SiO_2), Si nanoquantum dots (SNQD) and Si/SiO_2 superlattices.

In this paper we present a SiOx (x<2) thin film containing nanocrystalline silicon particles. Patterning was obtained exposing the nc-Si/SiO_2 sol-gel film to a Clark MXR CPA-2010 fs laser (387 nm) for optical modulation. The refractive index changes and optical properties were monitored with spectro-ellipsometric analyses.

Experimental

Nano-Si/SiO2 layers. The nc-Si/SiO_2 layers were deposited in sol-gel technology. The precursor for the SiO_2 solution was TEOS (TetraEthyl OrthoSilicate - $Si(OC_2H_5)_4$ from Sigma Aldrich). Ethanol was used as solvent. The solution containing silicon nanoparticles was added dropwise to the TEOS solution. This solution was then filtered through an Acrodisc syringe filter of 200 nm porosity (Sigma Aldrich) and deposited by spin-coating (200 rpm for 20 s) on a silicon wafer. After spin-coating the wafers were dried in air at 100 °C. A final thermal treatment at 400 °C was performed, resulting in amorphous SiO_2 films embedding the Si nanoparticles. A higher temperature thermal treatment in air is not recommended as it can lead to the oxidation of the nc-Si particles during the heating.

The nano silicon particles added to the sol-gel solutions were obtained through smooth anisotropic etching in a tetramethylammonium hydroxide (TMAH) 20 % solution in methanol at 50 °C. Traditional etching in KOH is not suitable for Si particles due to contamination with potassium ions. According to Biswas at al. [7] the etching rate for the conditions stated is 4 μm/h. At this rate, given the average particle size, the approximate etching time is 5-15 min (hydrochloric acid added to pH = 3). This solution, with nc-Si but also with particles of bigger size, is mixed with the TEOS solution, the larger particles being filtered away. The size distribution of the remaining nc-particles is then 5 - 200 nm (filter pore size).

Laser modified nano-Si embedded in SiO2 sol-gel layers.

The experiments were performed using a Clark-MXR CPA-2010 regenerative amplified Ti: sapphire laser system with a repetition rate of 1 KHz, pulse duration of 180 fs and a fundamental wavelength of 775 nm. UV 387 nm wavelength was produced by frequency doubling of the 775nm wavelength using a beta-BaB_2O_4 (BBO) crystal. The experimental setup is schematically shown in Fig. 1.

The beam diameter at the aperture of the laser system was 6mm. A high-energy variable diffractive attenuator was used for continuously adjusting pulse energy. The standard deviation in the pulse energy was ±1% of the average value. The nc-Si/SiO_2 film deposited on glass or on a silicon wafer is mounted on a computer controlled three axes stage (Aerotech). The beam was focused on the surface of the sample using a lens or an objective with different NAs. Pulse energies in the range of 0.5-1 μJ were used.

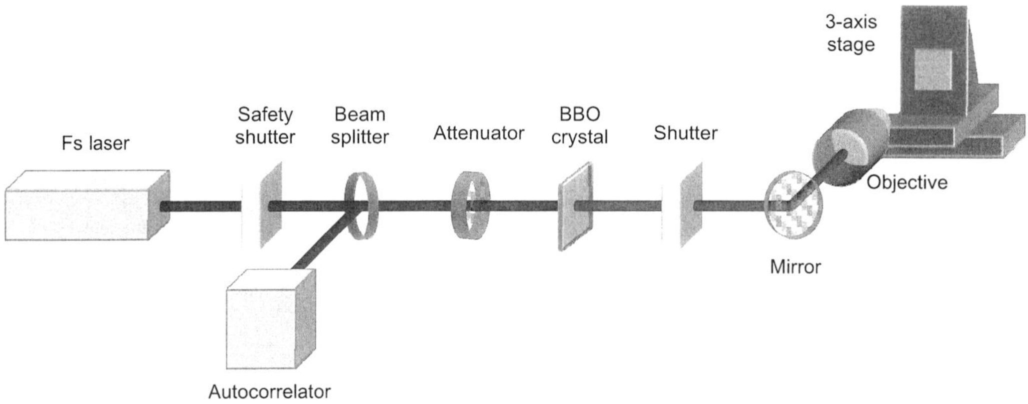

Fig 1 The experimental setup used for refractive index modulation using femtosecond UV laser technology.

Silica thin films on silicon and glass substrates were chosen for micro-machining as they are the most widely used structures in planar light-wave circuits. The typical modulation value of 10^{-3} can be dramatically increased due to nc-Si particles inclusion in the film. Buried waveguides were fabricated inside the partially thermally cured sol–gel film (Fig. 2). Irradiated films prepared by homogeneous irradiation of the surface were analyzed with spectro-ellipsometry. The laser-cured film depth is around 96 nm, starting from a sol-gel film with 100 nm thickness.

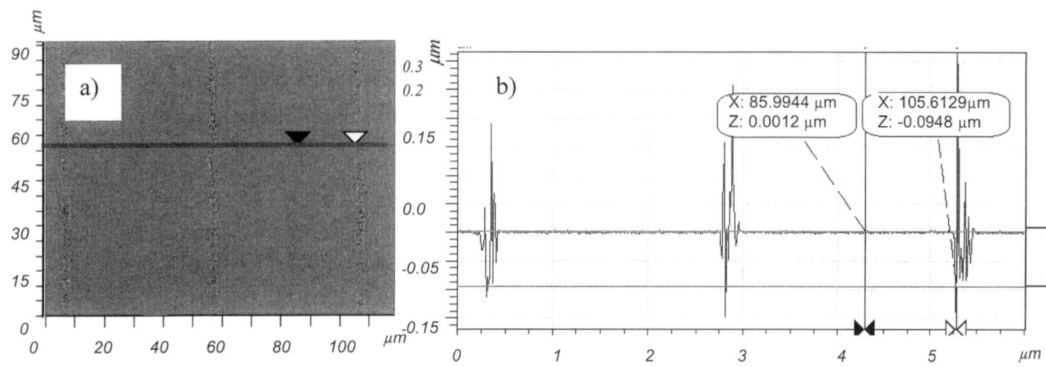

Fig 2 a) Buried waveguides fabricated using a 387 nm femtosecond laser on nc-Si/SiO_2 films deposited on silicon b) the optical profile of the surface

Results

AFM images were obtained using an Atomic Force Microscope (AFM) system with Veeco micromanipulation software. Fig. 3 shows the AFM image of the nano-silicon particles embedded into the SiO_2 sol-gel thin film. Nano-silicon clusters are present on the surface with varying sizes between 150-230 nm. The clusters are formed probably as a result of the spin coating process; however their presence is homogeneous on the surface.

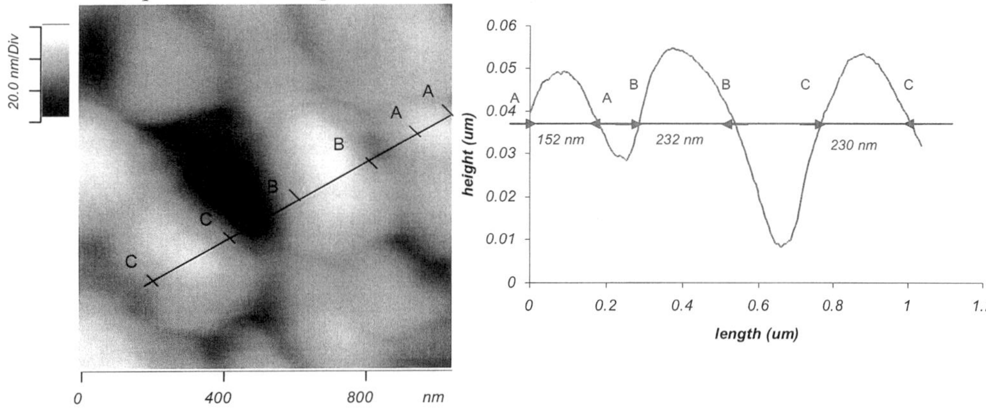

Fig 3 AFM images of a nc-Si/SiO_2 deposited on silicon wafer: clusters are seen at 20.0 nm/div with sizes between 150-230 nm

The polarized Raman spectra of the SiO_2 glass embedded with silicon nanoparticles were obtained with a high resolution Jobin-Yvon Horiba Raman confocal microscope using the 514.48 nm line of an Argon-ion laser as excitation source. The purpose of the measurement was to identify the nc-Si particles from the dried gel and to determine the effect of laser curing on the material. The experiment was performed on a nc-Si/SiO_2 dried gel (the drying performed in the air for a couple of weeks). In this way substrate influence is avoided, as well as that of oxidation of the nc-Si particles during thermal treatment (done with the purpose of removing organic residues from the thin film).

The spectral range investigated was 200–2000 cm^{-1} and the beam was consistently focused on the virgin sol-gel glass and separately on the laser-cured surface. A number of bands in the Raman spectrum of the sol-gel glass (fig. 4) are identified as belonging to bulk vitreous silica. Hence, certain aspects of the vibrational structure of the nc-Si/SiO_2 gel already resemble those of fused silica.

The main feature of the spectrum of fused silica is the broad band centered at ~440 cm^{-1} (ω_1) previously attributed to the Si-O-Si bond rocking and bending in SiO_4 tetrahedra [8]. This appears in both sol-gel spectra (fig. 4). The peak intensity on the irradiated glass is higher due to the flash heating process and quenching of the amorphous silicon during the laser curing [9].

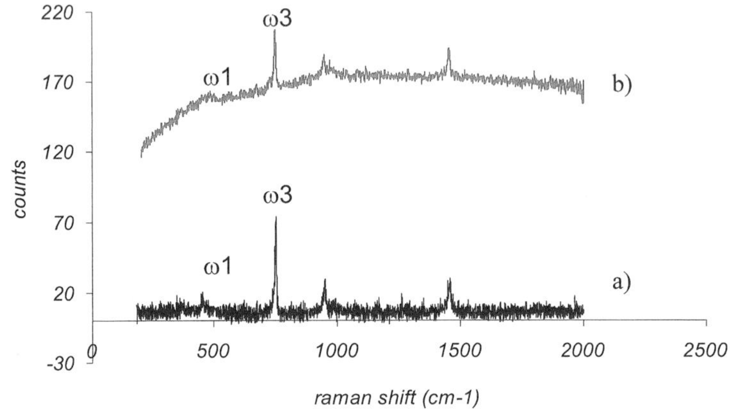

Fig 4 Raman spectrum of the a) unmodified nc-Si/SiO_2 gel and b) laser modified nc-Si/SiO_2 gel

The ω_3 peak is probably ascribed to a markedly broad overtone band centered at about 790 cm^{-1} attributed to the Si-O-Si symmetric stretching mode appears with high intensity in the uncured gel spectrum shifted at lower frequencies. The lower intensity of the peak in the cured gel case suggests changes into the silica glass structure during the laser curing with inclusion of nc-Si particles and new rearrangements into the network.

Other bands specific to the sol-gel spectra are observed at 950 cm^{-1} (typical of the Si-O stretching vibration of silanol (Si-OH) groups [10]) and at 1459 cm^{-1} (related with the CH_3 asym def [11]).We observe a progressive decrease of the intensity of the silanol band (Si–OH around 980 cm^{-1}) with respect to the intensity of the Si–O–Si band around 430 cm^{-1}. This corresponds to the condensation reaction between silanol groups to form the siloxane backbone. The silanol band is thus a very good indicator of the densification state of the gel [12]. It has been reported that the intensity of the Raman signal drops sharply when nc-Si concentration becomes less then 28 % and at the Si-poor end it is not detected, implying a sharp decrease in the number of silicon nanocristallites [13], explaining thus why the 520 cm^{-1} peak specific to crystalline Si is absent in the sol-gel glass Raman spectrum.

The optical properties (complex refractive indexes), thickness and the anisotropy of the sol-gel material were determined from spectroellipsometric (SE) data analysis with the aim to model its non-linear behavior and investigate its intrinsic non-linear properties. SE measurements were performed by using a spectroscopic ellipsometer (Jobin Yvon, Model UVISEL NIR). Ellipsometric data was acquired at an angle of incidence of 45° in the range 350 – 1600 nm with a step of 5 nm. The SE spectra present (Is, Ic) variables functions (fig 5a) of the (Ψ, Δ) ellipsometric angle measurements were Is and Ic equation are given by:

$$Is = \sin 2\Psi \sin\Delta$$

$$Ic = \sin 2\Psi \cos\Delta$$

A 3-layer model accurately describes the pristine nc-Si/SiO_2 film deposited on Si . The model takes into account the silicon substrate, a dense sol-gel SiO_2 film and a mixture of 50/50 void and film material. The optical model was described by the Tauc-Lorenz dispersion law in order to extract material characteristics. The film is highly transparent beyond 300 nm and adsorbent under 300 nm with a characteristic bandgap at 4 eV and a refractive index behavior similar to the one reported by

Chen at al [14]. The film is dense (a thin void layer of 5.95 nm being determined) with a thickness of approximately 100 nm.

The columnar structure of the laser-cured surface introduces uniaxial optical properties (optical axis perpendicular to the surface) similar to the one displayed by porous silicon [16]

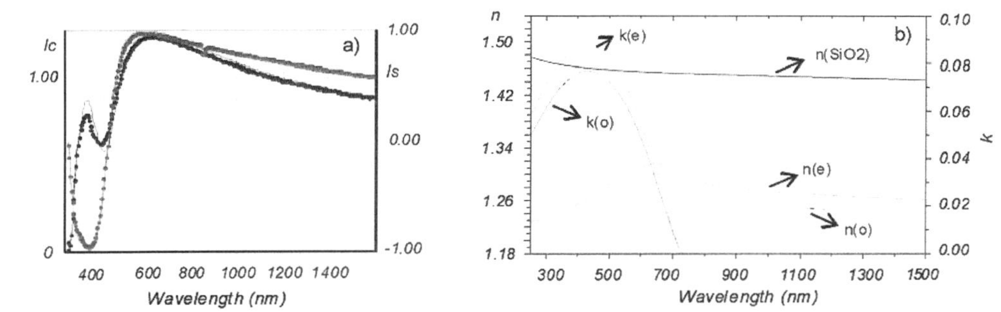

Fig 5 a) Ellipsometric data as (I_s, I_c), which relate to the ellipsometric angles Ψ and Δ; and afferent model (described in the text) fitted to the data; b) refractive index and coefficient of extinction dependence on wavelength. The 300-1600 nm data show low birefringence with ordinary n_o and k_o (perpendicular to surface) and extraordinary n_e and k_e (within the surface). The n(SiO_2) curve is for the pristine layer.

The 3-layer model (fig 5-a) takes into account the anisotropy of the structure determined by the silicon substrate and a mixture between 77 % SiO_2 and 23 % Si nanoparticles deposited on top of it. Fig. 5(b) shows the spectra of refractive indexes $n_{(o)}$ and $n_{(e)}$ – respectively extinction coefficients $k_{(O)}$ and $k_{(e)}$, in the range from 300 to 1600 nm, as well as their difference, obtained from the best values of all parameters. The index dispersion is that typical for dielectric materials. The measured birefringence of nc-Si/SiO_2 film is less than 0.031 in the whole UV–near IR range, respectively rather low. A positive birefringence is characteristic for the cured layer (n_e-n_o > 0) with the lower value close to 800 nm (corresponding to the highest value of the refractive index). In the extinction coefficient case the highest value corresponds to a maximum of absorption.

Table 1 Optical properties of laser-modified thin nc-Si/SiO_2 films on silicon and glass substrate

	nc-Si/SiO$_2$ sol-gel film on Si				**nc-Si/SiO$_2$ sol-gel film on glass**			
Film type	**Thickness [nm]**	**n at 300 nm**	**n at 1500 nm**	**Optical bandgap**	**Thickness [nm]**	**n at 300 nm**	**n at 1500 nm**	**bandgap**
Virgin sol-gel film	100	1.524	1.488	4	220	1.468	1.445	3.26
Laser cured sol-gel film	137	1.23	1.27	1.5	220	1.485	1.42	1.2
Refractive index change		0.294	0.218			-0.017	0.025	

The determined thickness for the cured film is 137 nm. The cured film is absorbent under 800 nm and the determined optical bandgap is 1.5 eV (table 1). The absorbent properties of the silicon substrate at the laser wavelength could influence the properties of the laser cured film (possible substrate melting with subsequent diffusion through the sol-gel layer). In this case, a thicker film was deposited on a glass substrate and the same experiment was performed. The refractive index change is lower but the optical bandgap is still the same (table 1) displaying a value closer to silicon bandgap. This change is possible if rearrangement inside the silica network takes place and Si nanoparticles become a part of the silica network (as concluded also by Raman measurements).

Conclusions

The patterned structures created by exposing a SiO_2 sol-gel film containing nano-silicon particles to a Clark MXR CPA-2010 fs laser (387 nm) display a very high refractive index variation (0.2 for silicon substrates, respectively 0.02 for glass substrates). The useful optical range of refractive index modulation is beyond 800 nm, respectively near-IR. The utility of the films is also below 800 nm – as absorbers, case in which the un-cured parts are VIS-transparent, while the laser-cured ones opaque. Applications of stated properties are holographic plates, optical isolators, etc, with a potential high-tech niche in phase matching for Mach-Zehnder interferometers in ultra-fast light pulse modulation applications – where active modulating elements cause static unbalance of the interferometer setup. Balancing the interferometer via electrical bias of the active elements reduces the available dynamic range of the modulator, hence a precise phase matching element in the optical circuit is very useful – especially one that can be produced without further material deposition, directly through irradiation. The optical transparency of the laser-cured material decreases drastically in visible (up to 600 nm) due to the 1.5 eV bandgap. This onset is correlated with the formation of a cured glassy phase (revealed by Raman spectroscopy) unaxial optically (determined ellipsometrically). The synthesis of laser-modified structures with a strong periodic modulation of refractive indices of the alternating structures was been developed with success. It is interesting to note that the pristine films display a moderately wide bandgap (3.87 eV), contrasting with that of the laser-cured ones – which can be useful in holographic and perhaps photovoltaic applications.

Acknowledgments

This research was carried out under the support of the EU Commission Grant MTKD-CT-2004-517165.

References

[1] J. Hubner, C.V. Poulsen, J.E. Pedersen Mogens, R. Poulsen, T. Feuchter,M. Kristensen, Proc. SPIE. 2695 (1996) 98.

[2] C. B. E. Gawith, G. D. Emmerson, S. P. Watts, V. Albani, M. Ibsen, R. B. Williams,P. G. R. Smith, S. G. McMeekin, J. R. Bonar, R. I. Laming, Lasers and Electro-Optics Europe 2003, CLEO/Europe 2003 Conference on 22–27 June 2003, p. 511.

[3] S. Nolte, M. Will, J. Burghoff, A. Tunnermann, J. Mod. Opt. **51** (2004) 16.

[4] J.W. Chan, T.R. Huser, S.H. Risbud, J.S. Hayden, D.M. Kroll, Appl. Phys. Lett. **82**(2003) 15.

[5] A. Saliminia, N.T. Nguyen, M.-C. Nadeau, S. Petit, S.L. Chin, R. Vallee, J. Appl.Phys. **93** (2003) 7.

[6] Yigang Li, Zian He, Hengsheng Tang, Liying Liu, Lei Xu , Wencheng Wang, Journal of Non-Crystalline Solids **354** (2008) 1216

[7] K. Biswas, S. Kal, Microelectronics Journal, 37 (2006) 519-525

[8] F.L. Galeener, A.E. Geissberger, Phys. Rev. B **27** (1983) 6199

[9] K. Awazu, Journal of Non-Crystalline Solids **337** (2004) 241–253

[10] E.I.Kamitsos, A.P. Patsis, Phys. Rev. B **48** (1993) 12 499

[11] Ying-Sing Li, Abdul Ba, Spectrochimica Acta Part A **70** (2008) 1013

[12] J.M. Nedelec, L. Courtheoux, at all, J Sol-Gel Sci Technol 46 (2008) 259

[13] T.V. Torchynska , A. Vivas Hernandez, Y. Goldstein, J. Jedrzejewskii, S. Jimenez Sandoval, Journal of Non-Crystalline Solids **352** (2006) 1152

[14] T.P. Chen, Y. Liu, M.S. Tse, O.K.Tan, P.F.Ho, K.Y. Liu, Phys. Rev. B **68**, 153301 (2003)

[15] A. Tanakaa, R. Saito, T. Kamikake, M. Imamura, H.Yasuda, Eur. Phys. J.D **43** (2007) 229

[16] M. Gaillet, M. Guendouz, M.Ben Salah, B. Le Jeune, G. Le Brun, Thin Solid Films, **455-456**(2004) 410

Solid State Phenomena Vol. 154 (2009) pp 107-112
online at http://www.scientific.net

Urethane Magnetorheological Elastomers – Manufacturing, Microstructure and Properties

Anna Boczkowska[1, a], Stefan F. Awietjan[1,b]

[1]Warsaw University of Technology, Faculty of Materials Science and Engineering, Woloska 141, 02-507 Warsaw, Poland

[a]abocz@meil.pw.edu.pl, [b]sawietjan@inmat.pw.edu.pl

Keywords: magnetorheological elastomers, carbonyl iron, microstructure, compressive strength, rheological properties.

Abstract. In this paper studies on urethane magnetorheological elastomers (MREs) microstructure in respect to their magnetic and mechanical properties are reported. MREs were obtained from a mixture of polyurethane gel and carbonyl-iron particles cured in a magnetic field of 100 and 300 mT. The amount of particles was varied from 1.5 to 33 vol. %. Samples with different arrangements of particles were produced. Effect of the amount of ferromagnetic particles and their arrangement on microstructure and properties in relation to the external magnetic field was investigated. The microstructure was studied using scanning electron microscopy. Magnetic properties were measured using vibrating sample magnetometer. Rheological and mechanical properties under compression were also examined.

Introduction

Magnetorheological elastomers (MREs) are materials with rheological properties which can be changed continuously, rapidly and reversibly by an applied magnetic field. They are the solid analogues of magnetorheological fluids (MRFs), consisting of magnetically permeable particles (such as iron) added to a viscoelastic polymeric material prior to crosslinking. A strong external magnetic field is applied before the curing process of the polymer [1-4]. The field induces dipole moments within the particles, which seek minimum energy states. Chains of particles with collinear dipole moments are formed and curing of the polymeric matrix material locks the chains in place [1]. In this orientation, the particles can form separate chains in three-dimensional simple lattice structures or even more complicated structures, where particles have multiple interaction points. This is the main difference between MRE and MRF structures which results in their elevated properties [5].

The advantage of MREs is that ferrous particles are not subject to sedimentation. The thermal stability of MREs is greater than for MRFs and their resistance to degradation is higher. The amount of particle filler can also be lower in MREs. As a result, the weight of sensing and actuating devices is reduced and due to the MRE microstructure, the time, following signal introduction, for a strain change response to applying a magnetic field is shortened [6]. MRFs have a field-responsive yield stress, whereas MREs have a field-responsive modulus. The elastic modulus of an MRE can change by 30-40% or more, under the influence of an applied magnetic field [4]. The relative change of modulus is higher for soft elastomers and the largest magnetorheological effect on modulus in MR solids reported so far is about 60% [5]. Interest in intelligent materials has increased significant recently and devices using rubber-like smart or active materials have motivated a number of advances in basic and applied research.

Elastomers represent an important class of engineering materials used for vibration and noise reduction, but controllable elastomeric materials based on magnetically conductive particles promise more functionality by varying properties interactively and therefore can play an important role in bringing modern control technologies and smart materials together [7-9].

MREs contain different magnetic particles (mainly carbonyl iron) in sizes from a few µm to a few hundreds of µm embedded in polymer matrices having low magnetic permeability. Pure iron has the highest saturation magnetisation of known elements, high permeability and low residual magnetisation, providing high short-term inter-particle attraction. MREs can find multiple applications in machine systems [8-10].

The results presented in this paper concern the studies on development of MREs based on ferromagnetic particles in a polyurethane matrix. The influence of the amount of the ferromagnetic particles and their arrangement in relation to the external magnetic field was investigated. Scanning electron microscopy was used to observe MRE microstructure. The particles orientation and their arrangement were also investigated with vibrating sample magnetometer (VSM). Rheological properties of MREs were studied with and without application of an external magnetic field, as well as their compressive strength.

Materials and measurement methods

Urethane magnetorheological elastomers were synthesized from polyether polyol VORALUX® HF 505 used in a blend with 14922 polyol and isocyanate compound HB 6013, supplied by Dow Chemical Company. Urethane elastomer obtained is characterized by relatively low density, hardness and stiffness. Its low viscosity before curing during MRE processing allows particles to arrange easily into aligned chains [11].

The ferromagnetic component used for the MREs fabrication, was carbonyl–iron powder with particle size ranging from 6 to 9µm, produced by Fluka. The amount of the carbonyl iron particles was varied from 1,5 to 33,0 vol. %.

The samples were produced with randomly dispersed and aligned carbonyl iron particles. They were subjected to a magnetic field of 0,1 and 0,3 T during curing to produce aligned carbonyl chains within the elastomer.

The microstructure of brittle fracture surfaces of MREs was observed by scanning electron microscopy (SEM) LEO 1530 Zeiss. Magnetic properties were studied by Lake Shore vibrating sample magnetometer (VSM). The compression tests were performed using MTS QTest/10 machine with magnetic coil device. Cylindrical samples with 20mm in diameter and 25mm in height were compressed at 5mm/min speed without and with magnetic field up to 0.3T. Rheological properties were evaluated with the application of Ares Rheometer from TA Instruments (plate-plate system, plate diameter - 20 mm, gap - 2 mm, magnetic field range - 0÷0,6 T). Experiments were conducted at 25°C on samples with various content of particles. The chains of particles within the samples were oriented at 0, 90 or 45° to the magnetic field lines.

Results

Observations of brittle fractures microstructure of MREs by SEM were carried out to find the arrangement of particles. As it is shown in Fig.1A the existence of aligned chains of particles in the elastomer matrix was found when curing process is conducted under the magnetic field. The existence of structural and magnetic anisotropy was confirmed by VSM studies. Tests were carried out parallel and perpendicular to the sample long axis, corresponding to the magnetic field direction during curing. In Fig. 1B an example of hysteresis loops obtained for the MREs with different carbonyl-iron volume fractions is shown.

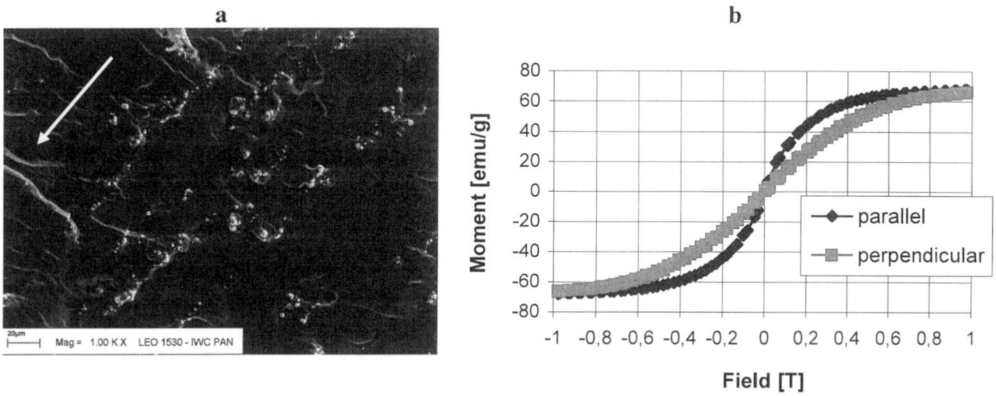

Fig.1 SEM image (a) and magnetic properties (b) of MREs with 11,5vol.% Fe, cured under magnetic field strength of 0,3T. White arrow shows magnetic field direction during curing.

It was found that the alignment of the chains of particles depends on the amount of the particles within the matrix. For the samples with particles content varying from 1,5 to 25 vol.% chains of particles were observed and hysteresis loops proved structural and magnetic anisotropy. SEM observations and hysteresis loops indicate also that for the highest ferrous particles content (33 vol. %) an isotropic network of particles is formed. It leads to the conclusion that the particles can arrange themselves easily into chains only for lower volume contents. High carbonyl-iron volume content (33 vol. %) in the PU matrix leads to the formation of more complex microstructures similar to the three-dimensional matrices.

The anisotropy coefficient (A_b), calculated from the hysteresis loops, expressed by the ratio of magnetic moments measured at a magnetic field strength of 0,2 T, respectively parallel and perpendicular to the alignment direction, for different carbonyl-iron particle volume content and magnetic field strength applied during curing is shown in Fig. 2.

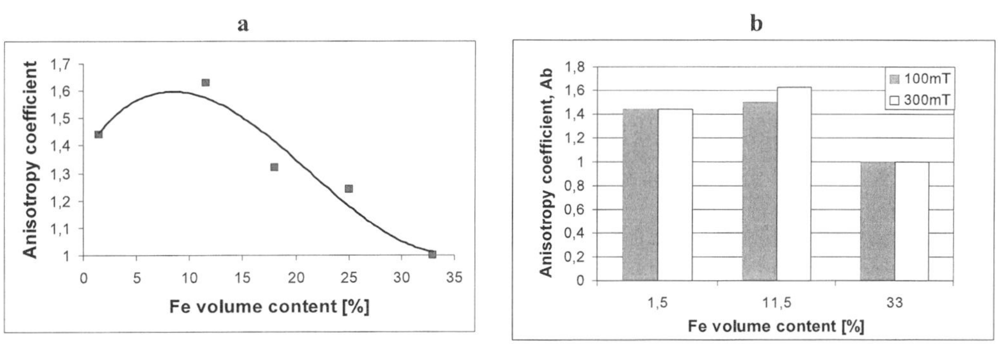

Fig. 2 Comparison of anisotropy coefficient for MRE with different amount of carbonyl-iron, cured at 0,3T (a) and under magnetic field strength of 0,1 and 0,3T (b).

It was found that there is a maximum value of the anisotropy coefficient corresponding to 11,5 vol.% of carbonyl-iron. The anisotropy coefficient decreases with further increase in the volume fraction of particles. Moreover, for the sample with the highest A_b a significant effect of the aligning magnetic field is visible; higher magnetic field leads to a better alignment.

Structural and magnetic anisotropy has not been found in the specimens having 33 vol. % of carbonyl-iron, therefore, the anisotropy coefficient is equal to 1 for these specimens. The result can be attributed to the formation of the network of particles, in contrast to the chains of particles that characterize anisotropic composites. This was also observed on SEM images.

In Fig. 3 the results of compression tests of samples with different amounts of iron powder cured under 0.3 T magnetic field are shown. They were compressed without and within 0.3 T magnetic field applied. It can be seen that under external magnetic field samples became stiffer and they are distinguished by higher compressive strength. This effect is stronger with the rise of the ferromagnetic particles amount in the material. For the samples with the smallest amount of the particles the stress-strain curves are almost overlapped.

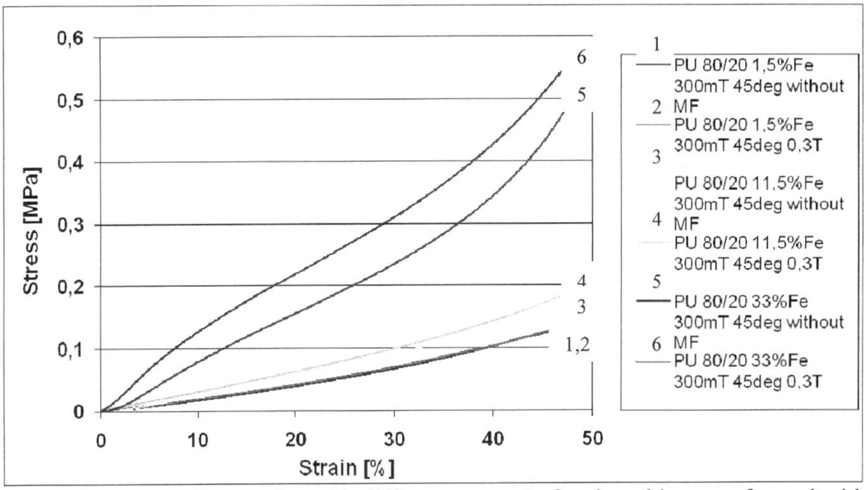

Fig. 3 Compressive curves for samples with different amount of carbonyl iron, performed with 0.3T and without external magnetic field.

The MREs are intended to be used as structural materials in applications where the load is often of a dynamic type. In cyclic dynamic loading, the material deforms and returns back to its original form during one cycle. The oscillating force is varied periodically, usually with a sinusoidal amplitude at the angular frequency ω.

In viscoelastic materials, some of the deformation energy input is stored and recovered during each cycle and some is dissipated as heat. The storage modulus, G' represents the ability of the viscoelastic material to store the energy of deformation, which contributes to the material stiffness. The loss modulus, G" represents the ability of the material to dissipate the energy of deformation. In this study, the elastic (storage) modulus G' was measured as a function of angular frequency ω.

As it is known from SEM and magnetic studies, the highest anisotropy coefficient was recorded for the sample with 11,5 vol. % of ferromagnetic particles. Therefore, the rheological properties of such MREs were compared to MREs with 33 vol. % Fe, which is characterized by a three-dimensional network of particles (Fig. 4). Investigations were carried out with and without application of an external magnetic field of 0,1 T and 0,2 T.

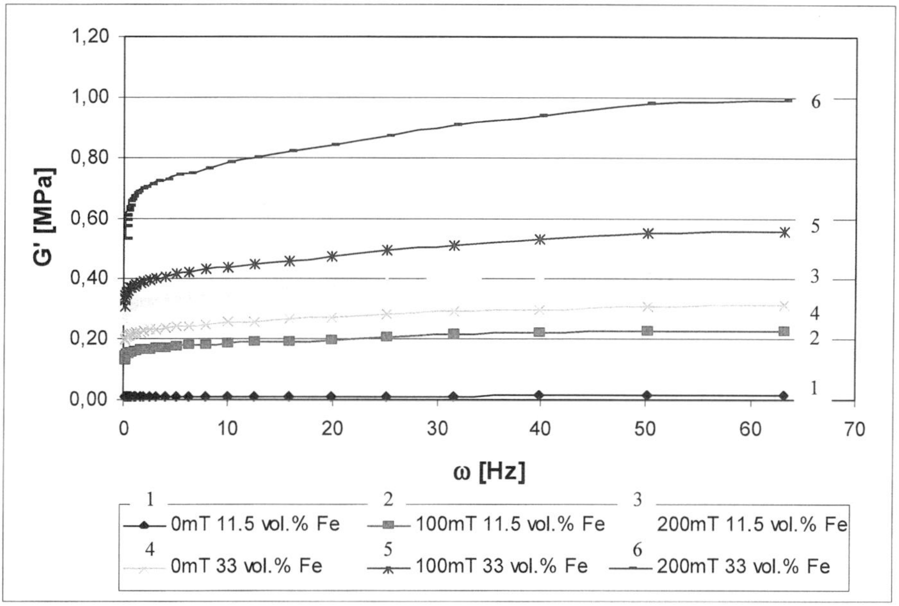

Fig. 4 Storage modulus G' versus frequency ω for MREs with 11,5 and 33 vol.% Fe, cured under magnetic field of 0,3T. Chains of particles are oriented parallel to the applied magnetic field.

As shown in Fig. 4, the value of G' increases with the increasing of the magnetic field for MREs with 11,5 and 33 vol. % Fe. Due to the higher amount of ferromagnetic particles in the PU matrix, higher values of G' for MRE with 33 vol. % Fe are observed. To compare changes in elastic modulus under external magnetic field the relative change of storage modulus has been calculated (Table 1).

Table 1 Relative changes of storage modulus for 10 Hz frequency, $\Delta G'_{200mT} = G'_{200mT}-G'_0$.

Fe vol. %	MF during curing [mT]	$\frac{\Delta G'_{200mT}}{G'_0}$
11,5	300	2,95
33	300	1,83

As shown in Table 1 higher relative changes in G' modulus have been obtained for the sample with 11,5 vol. % Fe. This can be correlated with the highest magnetic anisotropy coefficient obtained for this sample.

Summary

It was found that the microstructure of the MRE depends on the amount of ferrous particles and manufacturing conditions. The orientation of the iron particles into aligned chains is possible for lower volume content of the ferromagnetic fillers. High carbonyl-iron volume content in the matrix leads to the formation of more complex three-dimensional lattices. Also the magnetic measurements confirmed the existence of the microstructure anisotropy for the lower volume content of the iron particles. The structural and magnetic anisotropy has not been found in the MREs with 33 vol. % of particles. To evaluate the effect of the external magnetic field on the magnetorheological properties compressive strength, storage modulus was measured. Both, the content of particles and their arrangement have significant effect on the properties of magnetorheological urethane elastomers. Compression test results showed that under external magnetic field samples are distinguished by a higher compressive strength. The course of the stress-strain curves is the highest for samples with 33 vol. % of Fe.

Rheological properties of magnetorheological urethane elastomers also depend on content of particles and their arrangement. Application of an external magnetic field leads to a significant increase in elastic modulus. Relative changes of storage modulus, calculated from obtained curves, show that the microstructure of the samples has a significant effect on their magnetorheological effect. Magnetorheological effect, expressed by relative change of storage modulus under magnetic field, is the highest for the sample with the highest magnetic and structural anisotropy.

The aligned particle network structure has a significant influence on the elastic properties of the composite material. Inside the chains the effective filler content is higher than the average filler content. By optimizing the content of particles and alignment, either the stiffness or the damping properties of MREs can be increased by applying a magnetic field.

Acknowledgements

This study is financed as a Targeted Research Project from funds of the Polish Ministry of Science and Higher Education within years 2006-2008.

References

[1] G.Y. Zhou: Smart Materials and Structures Vol. 12 (2003), p. 139
[2] M. Farshad, A. Benine: Polymer Testing Vol. 23 (2004), p. 347
[3] M.R. Jolly, J.D. Carlson, B.C Munoz, T.A. Bullions: Journal of Intelligent Material Systems and Structures Vol. 7 (1996), p. 613
[4] H.T. Banks, G.A. Pinter, L.K. Potter, M.J. Gaitens, L.C. Yanyo: Journal of Intelligent Material Systems and Structures Vol. 10 (1999), p.
[5] M. Lokander, T. Reitberger, B. Stenberg: Polymer Degradation and Stability Vol. 86 (2004), p. 467
[6] J.D. Carlson, M.R. Jolly: Mechatronics Vol. 10 (2000), p. 555
[7] M. Khoo, C. Liu: Sensors and Actuators A: Physical Vol. 89 (2001), p. 259
[8] N. Triantafyllidis, S. Kankanala: Journal of the Mechanics and Physics of Solids Vol. 52 (2004), p. 2869
[9] Y. An, M.T. Shaw: Smart Materials and Structures Vol. 12 (2003), p. 157
[10] M. Farshad, M. Le Roux: Polymer Testing Vol. 24 (2005), p. 163
[11] A. Boczkowska, S. Awietjan, K. Babski, R. Wróblewski, M. Leonowicz, in: *Smart Structures and Materials 2006: Active Materials: Behavior and Mechanics,* edited by William D. Armstrong, Proc. of SPIE vol.6170, 6170_28, (2006)

Solid State Phenomena Vol. 154 (2009) pp 113-120
online at http://www.scientific.net

Experimental Validation of Numerical Methods of MRE Simulations

Wiesław Szymczyk[1], Anna Boczkowska[2], Tadeusz Niezgoda[1], Konrad Zubko[3]

[1]*Military University of Technology, Faculty of General Mechanics*
Kaliskiego Str. 2, 00-908 Warsaw, Poland, e-mail: w.szymczyk@wme.wat.edu.pl
[2]*Warsaw University of Technology, Faculty of Materials Science and Engineering,*
Woloska 141, 02-507 Warsaw, Poland
[3]*Military University of Technology, Faculty of New Technologies and Chemistry,*
Kaliskiego Str. 2, 00-908 Warsaw, Poland, e-mail: kzubko@wat.edu.pl

Keywords: magnetorheological elastomer, FEM modeling, magneto mechanical coupling

Abstract. This paper deals with the development of magnetoactive elastomers (MREs) based on the carbonyl iron particles-filled polyurethane resin. Their stiffness can be changed easily by magnetic field. Such a property can be useful in construction of active vibration damping structural elements. For the needs of numerical modelling methods validation the elementary case of the two magnetic particles was investigated experimentally.

Special “macro samples” were prepared with pairs of ferromagnetic particles of spherical shape of diameter of 12.7 mm. They provided easy observations and measurements. The gap distance between particles was established on the level of ¼ of the diameter.

After application of the magnetic field particles started to attract each other like magnetic dipoles.

The mutual displacement of the dipoles was recorded in function of the magnetic field intensity, which was varied in the range100÷300 [mT].

The deformation field was also obtained from the digital image processing (DIC).

Then the experiment was simulated numerically with the use of the 3D FEM models. The dipoles were loaded by forces which were increased gradually until displacements reached values that were measured experimentally. Calculations were performed on the MSC Patran-MARC platform. The Neo-Hookean material model was used to describe properties of the resin matrix. Magneto-mechanical coupling was taken into consideration with the use of an iterative method.

The results of calculations were compared with the experimental results. The validation of the base modelling concept was successfully completed.

Introduction

Magnetorheological elastomers (MREs) fall into the category of smart materials. Properties of MREs can significantly change and vary in a wide range accordingly to the change in applied magnetic field. The increase in the magnetic field intensity results in rapid, continuous and reversible increase of the elastic modulus.

MREs contain carbonyl iron particles of diameter of several microns each. At the beginning of the fabrication process they are homogenously dispersed in the liquid resin. Then the forming magnetic field is applied. This causes that the particles tend to group and form structures like chains. After polymerisation of the resin the forming magnetic field is switched off. The chains remain in the polymerised resin and are responsible for the unique behaviour of the material under applied work magnetic field. Particles in chains start to behave like magnetic dipoles and attract each other. Distances between particles change. This induces complicated distribution of strain and stress in the resin. The consequence is that stiffness of the material increases [1].

For a relatively high-modulus matrix such as natural rubber, the fractional modulus increase with the magnetic field can exceed 50%, while it can be even larger for low-modulus host materials. Thanks to this property MREs are used in constructions like: controlled vibration dampers, absorbers (TVAs), clutches, actuators, stiffness tunable mounts and suspensions, variable impedance surfaces, automotive suspension bushings, valves, brakes, safety restraint systems, semi active control systems, building vibration isolation, etc..[2] Performance of these systems can be easily controlled by the change of applied magnetic field [3].

Most of models of MRE material behaviour are based on the magnetic dipole interactions between adjacent particles. These interparticle interactions are then averaged over the entire sample to yield a model of the bulk magnetostrictive behavior. Magnetisable materials exhibit strongly nonlinear behaviours such as paramagnetism and ferromagnetism [4].

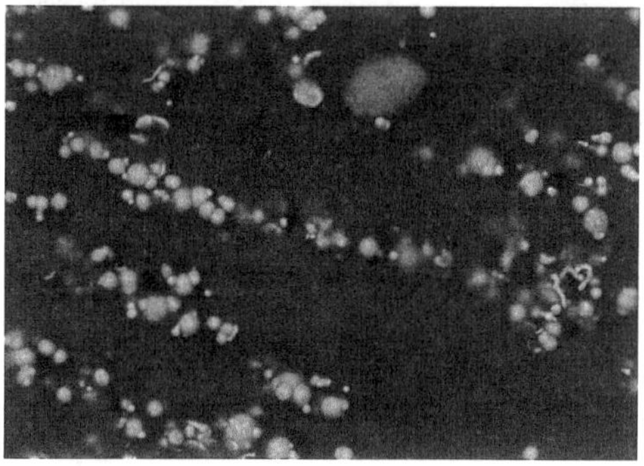

Fig. 1. Microphotograph of a MRE's structure: chains of carbonyl particles

Magneto mechanical coupling problem

Numerical calculations were performed on the MSC Patran-MARC platform. The Neo-Hookean material model was used to describe properties of the resin matrix. Magneto-mechanical coupling was taken into consideration with the use of an iterative method.

Magnetic and mechanical phenomena coupling in MREs acts as follows:

- placing a MRE in magnetic field induces forces between dipoles of carbonyl iron particles,
- these forces change the distances between the dipoles and try to change their orientation (when the directions of chains and direction of magnetic field forces are different),
- the change of distances between the dipoles changes the induced magnetic forces,
- such a coupling acts until the state of equilibrium is reached.

Material properties

The following assumptions were taken into considerations for the needs of calculations of magnetic forces acting between the two dipoles:

1.26 E-6	[H/m]	- magnetic permeability of free space,
7860	[kg/m^3]	- density of iron,
0.2353	[Am^2]	- magnetic moment of a 1g of iron in the full saturation state,
10.0		- relative permeability of the iron,

Areas of the model that correspond to the polyurethane resin had properties adequate to the Neo-Hookean material model: density 1.03 [g/cm^3], C_{10} ratio 22069,2 [Pa].

Magnetic force between two dipoles

Influence of the magnet onto a single dipole is formulated as:

$$F_x = \mu \,\mathrm{grad}(B_x)\,. \tag{1}$$

where μ is the magnetic moment of a single dipole, B_x magnetic induction in an arbitrary distance from the magnet pole;

Magnetic induction in an arbitrary distance from the magnet pole is described as:

$$B_x = \frac{b}{\sqrt{b^2 + x^2}} B_0 . \tag{2}$$

where B_0 is the magnetic induction generated by the single electromagnet, b is the width of the magnet pole. The gradient of the induced field is:

$$\text{grad}(B_x) = \frac{-b\,x}{(b^2 + x^2)^{3/2}} B_0 . \tag{3}$$

The mass of the dipole is given as:

$$m_k = \rho_{Fe} V_k . \tag{4}$$

Where ρ_{Fe} is the iron density and V_k the volume of the dipole.

Magnetic moment for the single gram of substance is given as:

$$\mu_{[g]} = 0.2353 \frac{\text{J}}{\text{T} \cdot \text{g}} . \tag{5}$$

From (4) and (5) we can obtain magnetic moment for the single dipole:

$$\mu = \mu_{[g]} m_k . \tag{6}$$

Experiments with the use of macro samples

An experimental campaign was carried out in order to calibrate the numerical model . It helped to verify the correctness of basic methods of FE modelling of MREs especially applied material model of the resin, method of estimation of magnetic forces between particles neighbouring in chains which are assumed to be aligned accordingly to the lines of magnetic forces.

An advanced experimental work is presented in [5]. Authors were able to fabricate MRE samples containing nano particles of various shapes ordered in chains.

In the paper another way of experimental investigations is presented, similar to the method described in [6] where so called Mullin's effect was studied with the use of macro samples with two neighboring spherical particles.

In the paper also "macro samples" were used . It enabled easy measurements of displacements of the "dipoles" in the magnetic field. The "dipoles" were spheres (steel balls) of diameter of 12.7 [mm] (Fig. 2).

Measurements of the displacements between the dipoles were conducted in the magnetic field which intensity was set to 100, 200 and 300 [mT].

The gap between dipoles was ¼ of the distance between their centers. The gap increase was measured with the use of equipment and software for digital image correlation (DIC). For this purpose light reflecting particles were introduced into the resin (Fig. 3.).

Fig.2. Macro sample was the resin plate with 2 steel balls simulating the dipoles

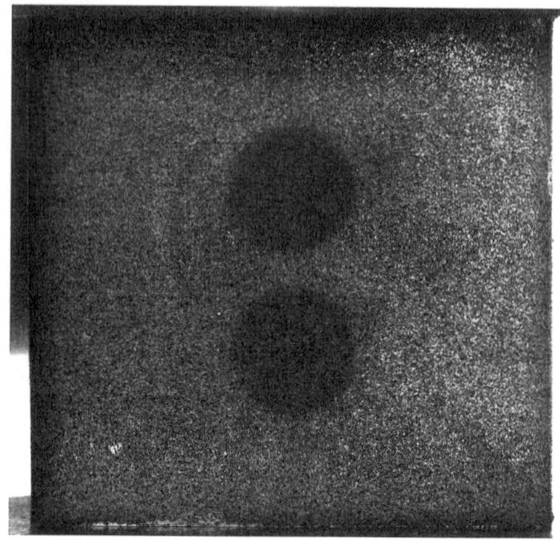
Fig. 3. Macro sample prepared for measurements with the use of DIC. The dipoles were aligned accordingly to the lines of magnetic field induction.

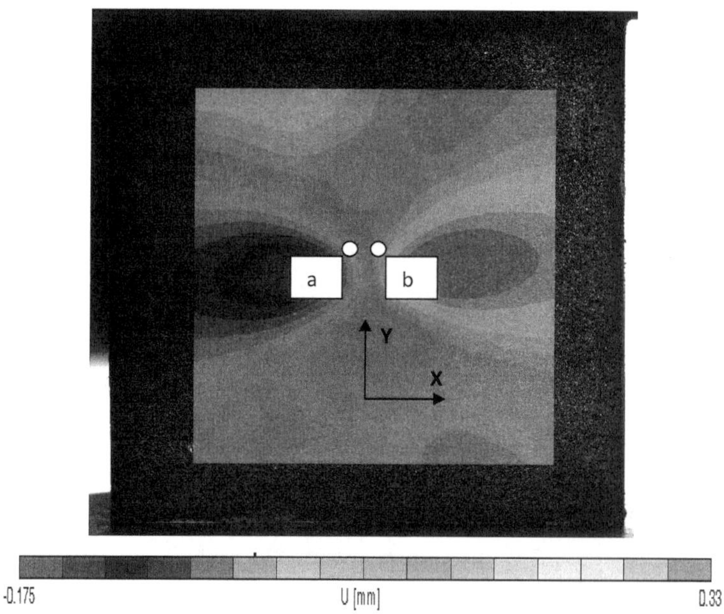

Fig. 4. Contour plot of horizontal component of displacements obtained from DIC [mm] for the magnetic field intensity 300[mT] – a) -0,17 mm b) 0,17 mm (on the external surface)

The accuracy of the DIC measurement of displacements was 0.05 [mm]. The decrease of the distance between the balls was used as the kinematic load in the numerical model.

Numerical simulations – model and results

Numerical model was built of 3D solid finite elements. Due to the symmetry of the macro sample and attraction between spherical particles only the 1/8 of the real macro sample was modeled (Fig. 5).

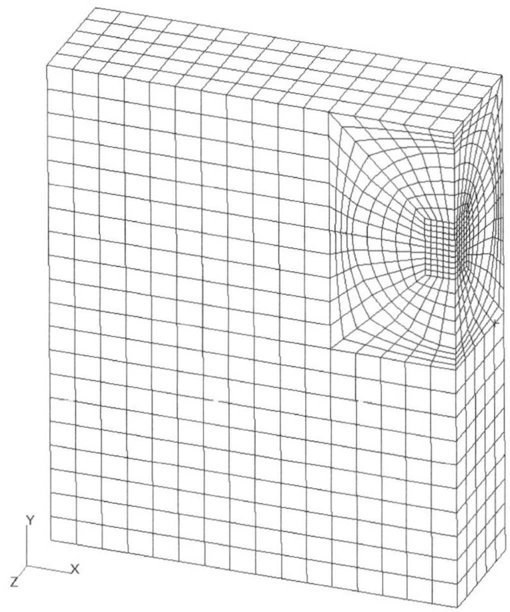

Fig. 5. 3D numerical model – mesh (the 1/8 of the macro sample), view onto planes of symmetry

The model was loaded by forces acting between the steel balls. Their values were changed until the resulting distance between the model balls reached this measured experimentally. Example results for the magnetic field intensity 300 [mT] are presented in figures 6-9.

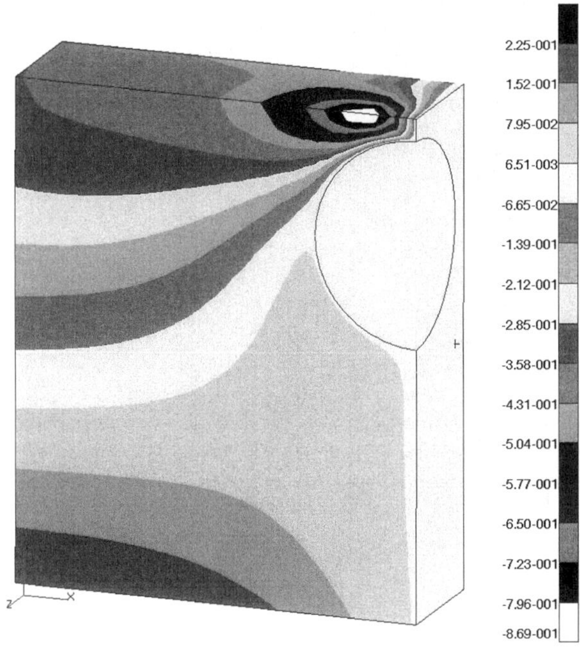

Fig. 6. Contour plot of x component of displacements [mm] – view onto cutting planes

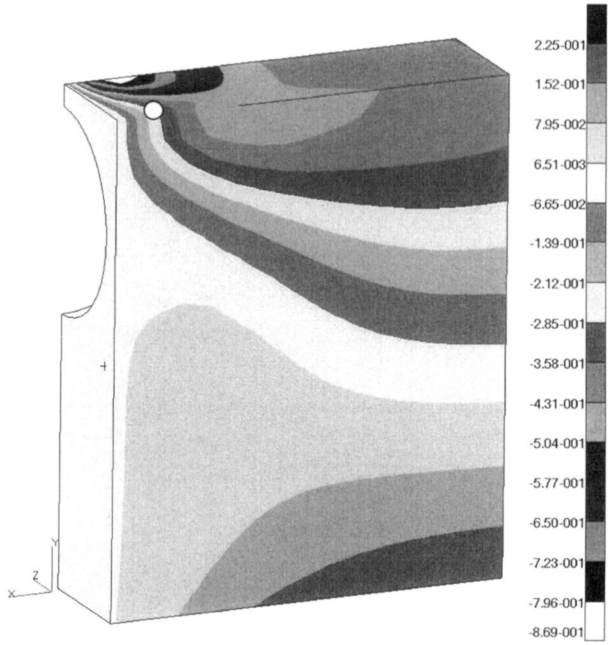

Fig. 7. Contour plot of x component of displacements [mm] – view onto external plane. For the appointed area values of displacement into x direction are between -0,14 and -0,21 [mm] (compare with figure 4)

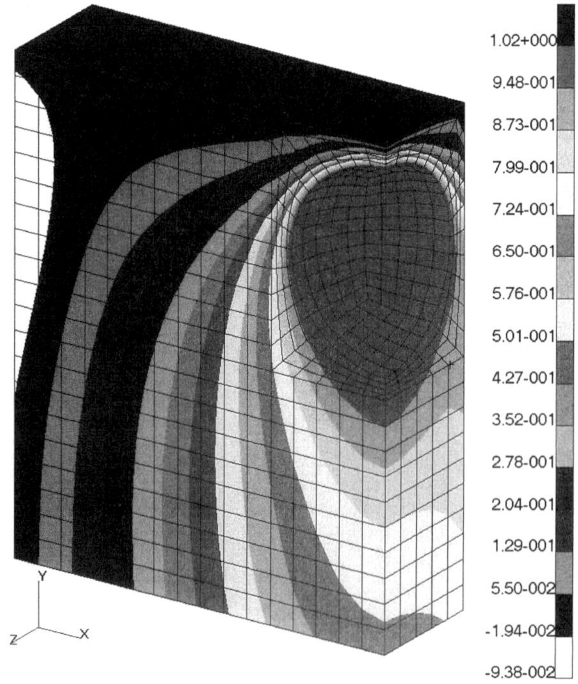

Fig. 8. Contour plot of y component of displacement [mm] – view onto cutting planes

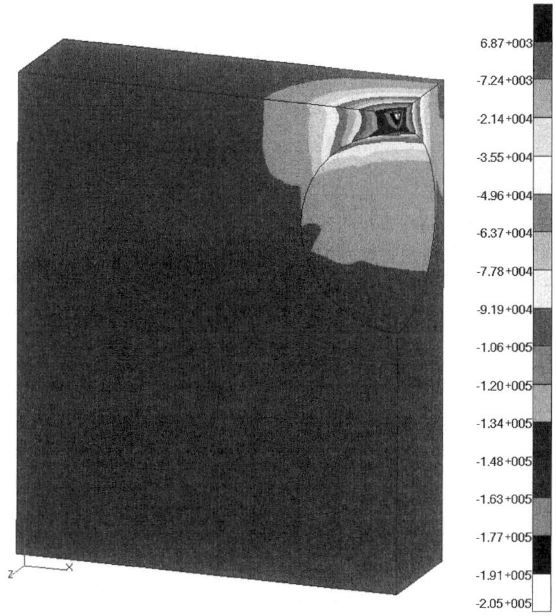

Fig. 9. Contour plot of y component of stress [Pa] – view onto external plane

Contours of displacements are similar in shape and values to those registered in the experiment. From comparison of figures 7 and 4 is visible that the value of x component of displacement obtained from the experimental measurements (0,17 [mm]) for the same location corresponds to the range calculate numerically (0,14÷0,21 [mm]). This allows to estimate magnetic forces acting between the balls in the macro samples and assume that distributions of stresses in the resin should be also in a good agreement with the real state (Fig. 9).

Table 1. Experimental data and numerical results

Experiment		FEM model	
Magnetic field intensity B [mT]	Displacement of centres of dipoles in the macro sample [mm]	Displacement of the centres of dipoles in the macro sample [mm]	Magnetic force acting between the dipoles [N]
		0,20197	0,32
100	0,2203	0,22553	0,36
200	0,5618	0,55522	1,00
300	1,0504	1,02160	2,40

In the Table 1 the experimental data and numerical results are presented. The force induced by magnetic field was assigned.

Conclusions

The model described in the paper was built only for the needs of verifications of methods of modelling of MREs. It was restricted only to 2 dipoles. Thanks to that it was possible to build experimental models (macro samples) which provided easy observations and measurements of the displacements during applying the magnetic field. Such macro samples are predicted to be also used in photo-elastically experiments which can provide additional way of verification of FE models.

Another simplification was that the particles in the chain were assumed to be aligned accordingly to the lines of magnetic induction. This excluded from considerations an additional moment effect. In case the dipoles are not aligned accordingly to the lines of magnetic induction the induced moment forces the particles to achieve such an alignment. Such an effect investigations were predicted for separate experiments and publications.

Experimental results (displacements of the 2 dipoles relative to the magnetic field intensities 100, 200, 300 [mT]) were used for calibration of the numerical 3D model of the macro samples. The model dipoles were loaded by forces simulating interaction between the dipoles. The values of the forces were increased until the resulting displacement were comparable to the experimental results.

The contour plots of the displacements obtained from calculations were comparable with those reached in the DIC experimental measurement due to their shape and values. The numerical model, material model choice were verified positively in such a manner. After that it is possible to accept numerical results that can be obtained from more complicated models which are comparable with the real microstructures. These models will be used to establish methods of effective properties of MREs assessment.

The simulation of magneto-mechanical coupling in the model that was used for the calculations on the MSC Patran-MARC model needed an external procedure. It was also impossible to simulate microstructures close to the real ones (Fig. 1). For such investigations the platform of ANSYS is more convenient, which allows to simulate complete magnetic circuits and achieve appropriate coupling effects automatically by the numerical system itself. This system enables also to investigate real like microstructure models taking into considerations random order of particles and distributions of their shapes and sizes.

Akcnowledgments: this study is financed as a Targeted Research Project from funds of the Polish Ministry of Science and Higher Education within years 2006-2008.

References

[1] Gong X.L., Zhang X.Z., Zhang P.Q., Fabrication and characterization of isotropic magnetorheological elastomers, Polymer Testing 24 (2005) 669-676.

[2] Carlson, J., Jolly, M., 2000. MR fluid, foam and elastomer devices. Mechatronics 10, 555–569.

[3] Ginder, J.M., 1996. Rheology controlled by magnetic fields. Encyl. Appl. Phys. 16, 487–503.

[4] Ginder, J.M., Clark, S.M., Schlotter, W.F., Nichols, M.E., 2002. Magnetostrictive phenomena in magnetorheological elastomers. Int. J. Mod. Phys. B 16, 2412–2418.

[5] Imrea A., Csabaa G., Bernstein G.H., Poroda W., Metlushkob V., Investigation of shape-dependent switching of coupled nanomagnets, Superlattices and Microstructures 34 (2003) 513–518.

[6] Coquelle, E., Bossis, G., Szabo, D., Giulieri, F., in press. Micromechanical model of an elastomer filled with particles organized in chainlike structures. Journal of Materials Sciences.

Solid State Phenomena Vol. 154 (2009) pp 121-126
online at http://www.scientific.net

Elastomers containing fillers with magnetic properties

Marian Zaborski, Joanna Pietrasik, Marcin Masłowski
Institute of Polymer & Dye Technology, Technical University of Lodz,
Stefanowskiego 12/16, 90-924 Lodz, Poland

Corresponding author marian.zaborski@p.lodz.pl

Keywords: Iron oxides, Fillers, Rubber composites, Magnetic properties

Abstract: Composites of rubbers such as ethylene – propylene and acrylonitrile – butadiene with micro size Fe_3O_4, Fe_2O_3, and nano size Fe_3O_4 as the fillers were tested. Iron oxides were found to be active fillers which improved mechanical properties of elastomers. Additionally, they also changed their magnetic properties. Furthermore, the surface of micro size Fe_3O_4 was modified to upgrade mechanical properties of the vulcanizates. Itaconic acid, methacryloxypropyltrimethoxysilane and monododecyl maleate were used as modifying agents. To improve dispersion of applied fillers in polymers matrix, non-ionic (NF-PEG), anionic (calcium stearate), cationic (DTAB, DTAC, DDAB, CTAD) and ionic liquids ($HMIMBF_4$, $BMIMPF_6$, $OMIMPF_6$) were used. It was observed that both non-modified and modified micro iron oxides and nano iron oxides improved the properties of the vulcanizates the most effectively.

Introduction

Magnetorhological elastomers are a new class of smart materials whose rheological properties can be changed by applying an external magnetic field [1]. Usually they contain micro- or nanoparticles of iron compounds dispersed in elastomer matrix instead of silica oil as it is common in the case of magnetorheological liquids [2-4]. Spherical carbonyl iron (CI) microparticles are commonly used [5]. The size of magnetically polarizable particles has great influence on the stress of magnetorheological materials. It is well known from literature [6] that aggregation of magnetic particles results in local anisotropy and dispersion irregularity that drastically diminishes magnetical and magnetorheological effect of materials. Therefore, it is evident that the best magnethoreological effect can be achieved by dispersion stability improvement.
The objects of our studies are the composites of ethylene – propylene, acrylonitrile – butadiene and silicone rubber with micro and nano size iron oxides as the fillers. Although iron oxides are well known to enhance the oxidation process of natural rubber materials [7], we believe that the application of more resistant polymers such as acrylonitrile – butadiene, ethylene – propylene and silicone rubbers will deliver materials with sufficient durability while demonstrating good magnetic and improved mechanical properties.

Material and methods

Materials

Acrylonitrile – butadiene rubber (NBR) – Europrene N3960 was purchased from Lanxess, ethylene – propylene rubber (EPM) Dutral CO 054 was obtained from Montedison Ferrara. Crosslinking agents: dicumyl peroxide (DCP), triallyl-1,3,5-triazine-2,4,6(1H,3H,5H) (TAC), zinc oxide (ZnO), mercaptobenzothiazole (MBT) and sulphur (S) were purchased from Sigma Aldrich.

Composition of typical elastomer mixture: rubber - 100 phr, DCP - 2 phr, TAC - 0,5 phr, ZnO - 5 phr, MBT - 2 phr, S - 2 phr.
Fillers: magnetite (Fe_3O_4) with micro- and nano- size, iron nickel oxide ($NiFe_2O_4$), iron(III) oxide (Fe_2O_3) were obtained from Sigma Aldrich.
Modifying agents: itaconic acid (IA), monododecyl maleate (MDM) and methacryloxypropyltrimethoxysilane (MPTMS) were purchased from Sigma Aldrich.
Dispersing agents: dodecyltrimethylammonium bromide (DTAB), dodecylotrimethylammonium chloride (DTAC), hexadecyltrimethylammonium bromide (CTAB), didecyldimethylammonium bromide (DDAB), 1-hexyl-3-methylimidazolium tetrafluoroborate ($HMIMBF_4$), 1-butyl-3-methylimidazolium hexafluorophosphate ($BMIMPF_6$), 1-octyl-3methylimidazolium hexafluorophosphate ($OMIMPF_6$) were provided from Fluka, octadecylamine (ODA) was supplied from Riedel de Haën, nonylphenylpoli (ethylene glycol) ether (NF-PEG) was purchased from "Rokita" Chemical Plant.

Methods

Particle size, Zeta potential and viscosity in water and paraffine oil were measured with Zetasizernano S90 , Zetasizernano 2000 and Rheometricscientific RM500, respectivily.
The mechanical properties were estimated according to ISO 37 standards. Magnetic properties of the samples were measured with magnetometer VSM LakeShore 7410, in the range of H from 200 to -200 G.
Heat ageing process was performed with FD series dryer (Binder), according to PN-82/C-04216 standards. Ultraviolet ageing process was run with UV2000 equipment (Atlas).
In order to modify the surface of the filler (100 g) it was mixed with solvent (300 ml) and modifying agent (10 g) and placed in ultrasonic mixer for 1 h. Then, solvent was evaporated with rotorvap.
Cross-linking density of vulcanizates was calculated with the basis of rapid solvent-swelling measurements (in toluene and in toluene with ammonium fumes) from the Flory-Rehner Equation.
Elastomer mixtures, based on maintained rubbers, with iron oxides were prepared by the common manufacturing procedures of rubber. Iron oxides particles were mixed with the rubber together with vulcanizations system using two – roll mills at 30 °C. Then the mixtures were vulcanized at 160 °C for τ 0.9 (τ 0.9 - optimum vulcanization time).

Results and discussion

Table 1. The characteristics of iron oxides

Type of irone oxide	Aggregates size in water [nm]	Aggregates size in paraffine oil [nm]	φ/φ_0	Isoelectric point [mV]
micro (Fe_3O_4)	881	18100	11,3	9,1
nano (Fe_3O_4)	589	9300	1,12	9,3

φ – dynamic viscosity of suspension
φ_0 – viscosity of paraffine oil (0,172 Pa*s)

The surface of both oxides has the similar character since they both have the same isoelectric point. However, they demonstrate very different tendency toward agglomeration in water and in paraffine oil. That is reflected in the aggregates size and φ/φ_0 values. Nanoparticles exhibit great tendency toward aglomeration (in the contrary to microparticles), as a result viscosity shows lower values. (Tabel 1).

Table 2. Cross-linking density, mechanical and magnetic properties of EPM vulcanizates containing iron oxides

Type of irone oxide	phr	ν_e*10 [5]	SE_{100} [MPa]	T.S [MPa]	E.B [%]	Ms [emu/g]	Mr [emu/g]
-	0	12,6	0,99	1,37	301	-	-
micro Fe_3O_4	60	11,6	1,08	11,11	1024	31,92	3,43
nano Fe_3O_4	60	18,0	1,23	6,29	762	26,09	4,68
nano Fe_3O_4	120	16,2	1,44	3,5	1181	37,38	2,66

ν - cross-linking density
SE_{100} – modulus at 100% elongation
TS – tensile strength
EB – elongation at break
Ms – magnetization
Mr – retentivity

Only the nano size iron oxide influences the cross-linking density of vulcanizates (Table 2). However, the difference becomes smaller when the bigger amount of nano size iron oxide is added. Concerning the mechanical properties of the composites, it appears that the micro iron oxide is the most effective filler for EPM. Tensile strength increases 10 times upon addition of 60 phr of micro Fe_3O_4 in comparison to unfilled system. This is probably due to worse dispersion of nano oxides in the elastomer. Also modified micro size Fe_3O_4 (Table 3) allows to obtain composites with better mechanical properties than nano form. (Table 2).

Table 3. Cross-linking density, mechanical and magnetic properties of EPM vulcanizates containing modified micro size magnetite (Fe_3O_4 – 90 phr).

Type of irone oxide	ν_e*10 [5]	SE_{100} [MPa]	T.S [MPa]	E.B [%]	Ms [emu/g]	Mr [emu/g]
Fe_3O_4 non-modified	14,0	1,10	8,54	1549	40,06	3,79
Fe_3O_4 + IA	8,6	1,31	4,99	926	37,38	4,01
Fe_3O_4 + MPTMS	40,5	1,82	6,02	324	36,14	4,09
Fe_3O_4 + MDM	18,1	1,35	14,34	947	36,07	3,73

As expected, all composites exhibit magnetic properties. The more iron oxide is added to the rubber, the higher magnetization and retentivity is observed (Table 2,3,5,6,7). Moreover, modification of micro size Fe_3O_4 improves its magnetic characteristic. (Table 3).

Table 4. Cross-linking density, mechanical and magnetic properties of EPM vulcanizates containing micro size magnetite (Fe_3O_4) and silica A380 (20 phr).

Fe_3O_4 [phr]	v_e*10 ^{5}	SE_{100} [MPa]	T.S [MPa]	E.B [%]	Ms [emu/g]	Mr [emu/g]
-	12,6	0,99	1,37	301	-	-
20	24,6	2,02	18,33	800	10,91	1,63
40	28,6	1,64	17,89	857	19,98	2,92
60	26,9	1,71	15,79	932	25,47	3,70
80	26,7	1,67	11,85	1075	30,45	4,57
100	36,2	1,88	9,5	858	36,19	5,89

In order to enhance the mechanical properties of micro size iron oxide another approach was applied. Namely, silica was added together with Fe_3O_4. The highest values of tensile strength characterizes the vulcanizates with 40 phr of filler. On the other hand, with higher content of iron oxides, magnetization and retentivity increase. (Table 4).

Table 5. Cross-linking density, mechanical and magnetic properties of NBR vulcanizates containing micro size magnetite (Fe_3O_4).

Phr	v_e*10 ^{5}	SE_{100} [MPa]	T.S [MPa]	E.B [%]	Ms [emu/g]	Mr [emu/g]
20	12,0	1,18	16,70	761	2,14	15,8
40	15,4	1,89	12,31	693	3,69	25,3
60	18,5	1,98	13,30	776	5,33	32,8
80	32,6	2,55	17,74	760	6,06	37,1
100	36,1	3,22	16,09	717	8,90	56,4
120	52,2	3,97	15,27	702	7,42	46,6

Table 6. Cross-linking density, mechanical and magnetic properties of EPM vulcanizates containing nano size magnetite (Fe_3O_4 – 60 phr) and dispersing agents.

Type of dispersing agent	ν_e*10 [5]	SE_{100} [MPa]	T.S [MPa]	E.B [%]	Ms [emu/g]	Mr [emu/g]
-	18,0	1,23	6,29	762	26,09	4,68
DTAB	10,1	1,04	7,86	1023	24,38	3,91
DTAC	12,1	0,94	6,42	967	26,33	4,29
DDAB	12,8	0,97	7,41	906	25,75	4,34
CTAB	9,5	1,02	6,44	1055	25,48	4,12
$OMIMPF_6$	11,8	1,03	6,17	1065	24,66	3,80
NF-PEG	20,6	1,15	6,17	804	26,06	4,25
$BMIMPF_6$	16,0	1,11	8,72	927	25,62	4,09
Calcium stearate	29,9	1,05	8,92	904	28,47	4,67

In order to improve the dispersion of iron oxide properties, surfactants were added during the preparation of composites. Ionic, cationic, non-ionic dyspersants and ionic liquids were used. No matter what the character of the dispersant is, the mechanical properties of vulcanizates are slightly better than the properties of composite when no dispersant is used.

Table 7. Ageing factors of EPM vulcanizates containing modified and non-modified iron oxides

Type of irone oxide	phr	Heat ageing factor		UV ageing factor	
		$T.S^{st}/T.S$	$E.B^{st}/E.B$	$T.S^{st}/T.S$	$E.B^{st}/E.B$
-	60	1,06	1,45	-	-
Fe_3O_4 nano	60	1,25	1,08	-	-
Fe_3O_4 micro	60	0,92	0,96	-	-
Fe_3O_4 nano	120	-	-	0,93	0,92
Fe_3O_4 micro non-modified	90	-	-	0,87	0,99
Fe_3O_4 micro + IA	90	-	-	1,01	0,84
Fe_3O_4micro + MPTMS	90	-	-	0,88	0,86
Fe_3O_4micro + MDM	90	-	-	0,81	0,79

TS – tensile strength
EB – elongation at break
TS^{st} – tensile strength measured after ageing
EB^{st} – elongation at break measured after ageing

Typicaly metals with variable oxidation state can influence the mechanical properties of materials after their ageing. However, the conditions which were applied herein, for ageing, let us to maintain the values of tensile strengh and elongation at break (Table 7) of the EPM vulcanizates. As it can be seen from table 4, ageing factors are nearly the same for all types of tested composites.

Conclusions

Elastomers containing fillers with magnetic properties were prepared according to commonly used methods. Micro and nano size oxides and modified micro size Fe_3O_4 are active fillers of acrylonitrile – butadiene and ethylene – propylene rubber. They may be used as substances improving mechanical and magnetic properties. Dispersion of nano size iron oxides in the elastomer matrix can be improved by using dispersing agents.

References

[1] B.C. Munoz and M. R. Jolly, in: *Performance of Plastics,* edited by W. Brostow, Carl Haser Verlag, Munich (2001), p. 553.

[2] M. R. Jolly, J. D. Carlson and B. C. Munoz, Smart Mater. Struct. 5, 607 (1996).

[3] Y. Wang, Y. Hu, X. Gong, W. Jiang, P. Zhang and Z. Chen, J. Appl. Polym. Sci. 103, 3143 (2007).

[4] Y. Wang, Y. Hu, H. Deng, X. Gong, P. Zhang, W. Jiang and Z. Chen, Polym. Eng. Sci. 264, 154 (2006).

[5] W. I. Kordonski, S. R. Gorodkin and Z. A. Novikova, *The influence of ferroparticle concentration and size on fluid properties,* 6th International Conference on Electo-Rheological Fluids, Magneto-Rheological Suspensions and their Applications, World Scientific, Yonezawa, Japan 1997.

[6] M. Lokander and B. Stenberg, Polym. Test. 22, 677 (2003).

[7] M. Lokander T. Reitberger and B. Stenberg, Polym.Degrad. Stability 86, 467 (2007).

Solid State Phenomena Vol. 154 (2009) pp 127-132
online at http://www.scientific.net

Tensile Stress-Strain Relationships of Magnetorheological Fluids under Various Factors

S. A. Mazlan[1,a], A. Issa[1,b], H. A. Chowdhury[1,c] and A. G. Olabi[1,d]

[1]School of Mechanical and Manufacturing Engineering, Dublin City University, Glasnevin, Dublin 9, Ireland.

[a]saiful.mazlan2@mail.dcu.ie, [b]ahmed.issa2@mail.dcu.ie, [c]chowdhh2@mail.dcu.ie, [d]abdul.olabi@dcu.ie

Keywords: Stress-strain relationship, Magnetorheological Fluid, Squeeze Mode, Tensile Stress

Abstract. This paper presents experimental investigations of three MR fluid types under the influence of several factors in tension loading mode. One MR fluid was water-based, MRF-241ES, while the other two were both hydro-carbon based, MRF-132DG and MRF-122-2ED. The magnetic properties of the MR fluids varied significantly due to a higher particle density in the water-based MR fluid compared to the hydrocarbon-based MR fluids. Tension as a squeeze mode, is as an operational mode where two flat parallel surfaces, standing opposite to each other, are pulled apart from each other by an external force, acting along the path of the magnetic flux lines. The experiments were performed in a vertical direction in the environment of a DC magnetic field. Stress-strain curves of the MR fluids under tension showed similar characteristics despite the fact that different types of the carrier fluids were used. The results revealed that the magnitude of the stress, for a given strain value, depended on the applied current and the initial gap size. It was found that, for higher applied currents and smaller initial gap sizes, there were larger stress values. However, the tensile speed had no significant effect on the stress-strain curves of MR fluids.

Introduction

Magnetorheological (MR) fluids are commonly suspensions consisting of micron-sized magnetic particles and carrier liquids. MR fluids are currently enjoying renewed fundamental and applied research interests within the technical community. MR fluids are remarkable for their precipitous and their viscosity is capable of being reversible controlled by the applied magnetic field. In the presence of an applied magnetic field, the suspended particles appear to align or cluster and the fluid drastically thickens or gels. The flow resistance (apparent viscosity) of the fluid is intensified by the formed particle chain. When the magnetic field is removed, the particles return to their original condition, which lowers the viscosity of the fluid. The fluid structure is dependent on many factors such as volume fraction, magnetic field strength and carrier fluid. The fluid structure is also responsible for a rapid formation and reversible formation in either solid-like state or free-flowing liquid state. The changes of solid-liquid state or the consistency or yield strength of the MR fluid can be precisely and proportionally controlled by altering the strength of the applied magnetic field. These characteristics provide simple, quiet, rapid response interfaces between electronic control and mechanical systems [1].

Recently, despite the fact that much research work on MR fluids has been done, little applications have been developed. This is mainly due to inadequate understanding of the fundamentals and properties of MR fluids. Researchers carried out intensive investigations of the MR fluids behaviour under shear mode [2,3] and compression mode [4-6], but less attention has been given to the tensile behaviour. In order to understand tensile stress of MR fluids, the behaviour of ER fluids, which change their rheological behaviour in electric fields, can be adapted. Tian et al. [7] investigated the behaviour of ER fluids under different applied direct current voltages. The authors found that the tensile yield strength was proportional to the electric field for several values of electric field power and voltages. The tensile behaviour of ER fluids can be well approximated by normalized methods to address the electric field and the particle structure effects on the elongations [8]. Choi et al. [9] also observed that the tensile stress increased at higher values of the electric field. However, a thorough study on the tensile behaviour of MR fluids needs to be conducted. In this paper, experimental investigations have been carried out on three types of MR fluids under different applied current values, initial gap sizes and tensile speeds in tension mode.

Experiment

The MR fluids utilized in this study were commercially available, namely MRF-241ES, MRF-132DG and MRF-122-2ED, produced by Lord Corporation. Typical physical properties of these fluids are listed in Table 1. MRF-241ES was a water-based MR fluid, while MRF-132DG and MRF-122-2ED were hydrocarbon-based MR fluids. The fluids have the advantages of requiring low voltage power supplies in the range from 2 to 25 V, with current ranges from 1 to 2 Amps and power ratings from 2 to 50 watts [10].

Table 1. Physical properties of the tested MR fluids.

Property	Values / limits		
Codes of MR fluids	MRF-241ES	MRF-132DG	MRF-122-2ED
Base Fluid	Water	Hydrocarbon	Hydrocarbon
Operating Temperature [°C]	-10 to 70	-40 to 130	-40 to 130
Density Range [g/cm^3]	3.80 to 3.92	2.98 to 3.18	2.32 to 2.44
Solids Fraction [wt. %]	85	81	72

A schematic diagram of the experimental set-up is shown in Fig. 1 [4]. The experimental-set up consisted of five main components, namely Instron Machine, test rig, MR fluid, power supply and control computer. The test rig was placed in the testing area of the Instron Universal Electromechanical Testing (UTS) machine (model 4202). The communication with the machine was achieved via a built-in data acquisition board in the control computer. The Instron machine was operated in the vertical direction to obtain displacements and forces under tension modes. A measured amount of the MR fluid was filled in the gap between upper and lower cylinders, and surrounded by support cylinder. The current was supplied by the Xantrex$_{TM}$ XFR 150v, 18 Amps DC power supply. The magnetic field strength was adjusted by manually controlling the value of the current supplied to the electromagnetic coil. All experiments were carried out in a displacement control mode at a room temperature of 20 °C.

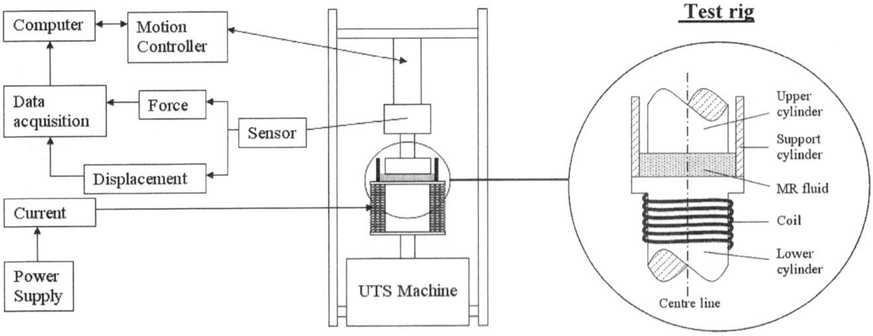

Fig. 1. Schematic diagram of experimental set-up.

Three sets of experiments have been accomplished. In all sets, the initial gap size between the two cylinders was adjusted to 1.0 and 2.0 mm. In the first set of experiments, the tension of the MR-241ES was achieved by pulling up the upper cylinder at various speeds of 0.5, 1.0, 1.5 and 2.0 mm/min. These experimental conditions were conducted with different applied currents of 0.4, 0.8, 1.2 and 1.6 Amps. In the second and third sets of experiments, the same parameters combinations and experimental procedure were applied as in the first set. However, the sample was changed to the MRF-132DG and MRF-122-2ED for the second and third sets respectively. During tension, the applied current was kept constant and the experiment was stopped when the final gap sizes reached about 75% of the initial gap size.

Results and Discussions

Generally, the MR fluid tension process, in terms of stress-strain relationship, can be separated into three regions [11] as shown in Fig. 2. The tensile stress curves in the first region increased dramatically to the peak point with increasing tensile strain. Subsequently, in the second region, the tensile stress curves declined rapidly for small values of the tensile strain. Finally, in the third region, the tensile stress curves further decreased tardily with further elongation.

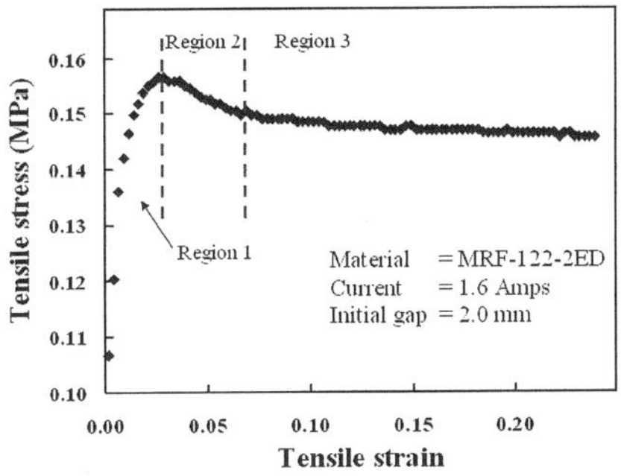

Fig. 2. Tensile stress-strain relationship of MR fluid [11].

The same assumptions, about the relative movement between the particles and the carrier fluid under the compression process [4] can be made to describe the behaviour of the MR fluids under the tension mode. As postulated by Tian et al. [7], the tension process is a relationship between structure strengthening effect and decreasing magnetic field strength. Structure strengthening effect is referring to particle chains reorganizing during the tension process, while decreasing magnetic field strength is referring to the gap expanding between the upper cylinder and the lower cylinder. During the tension process, in the first region, the structure strengthening effect was significant since the magnetic field strength was at the highest value. Both effects contributed to the highest tensile stress. After the peak point has been reached, in the second region, the magnetic field strength continued to decrease as the gap increased. Therefore, the tensile stress decreased drastically with decreasing the magnetic field strength. After tensile stress reached a certain point, the effect of structure strengthening could be considered as negligible. The effect of decreasing magnetic field strength was dominant in the third region. The tensile stress decreased slowly with further decrement of the magnetic field strength.

a. Effect of applied current

Fig. 3 shows the effect of the applied current on the MR fluids under a constant tensile speed of 1.5 mm/min and an initial gap size of 2.0 mm. As can be seen, under the same value of the tensile strain, the value of tensile stress was higher at higher values of applied currents. Lower values of tensile stress could be achieved as the applied current was reduced.

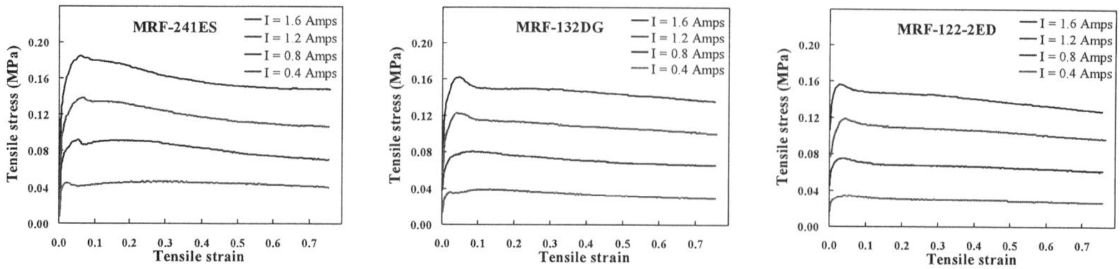

(a) (b) (c)

Fig. 3. Tensile stress versus tensile strain at a tensile speed of 1.5 mm/min and an initial gap size of 2.0 mm for (a) MRF-241ES, (b) MRF-132DG and (c) MRF-122-2ED.

Similar effect of the applied current could also be found in the discussion of the compression process in the preceding works [4-6]. Higher values of the tensile stress could be achieved by increasing the applied current [7,8]. When an external magnetic field was applied to the MR fluids, the polarized particles formed particle chains along the field direction. The interactions between the particles at high applied currents were much stronger than those at low applied currents. The required force to initiate moving up the upper cylinder was high. This force was necessary to break up the particle chains, which corresponded to the highest value of tensile stress, shown in Fig. 3. As the gap size was further increased, the broken particle chains aggregated to form columns. Even though the interactions between the particles in columns were expected to be stronger than particle chains, the increases in gap size reduced the magnetic field strength. The effect of decreasing the magnetic field was more significant than the interaction between the particles in the columns. Therefore, as the gap size increased, the tensile strength of the MR fluid decreased. Tensile stresses of the MR fluids were higher, at higher applied currents, than tensile stresses at lower applied currents. This was because the magnetic field strength was one of the factors that influenced the tensile stress as demonstrated by different values of applied currents.

b. Effect of initial gap size

Tensile stress-strain curves of the MR fluids at a constant tensile speed of 1.5 mm/min, applied current of 1.2 and 0.8 Amps, and different initial gap sizes are shown in Figs. 4 and 5.

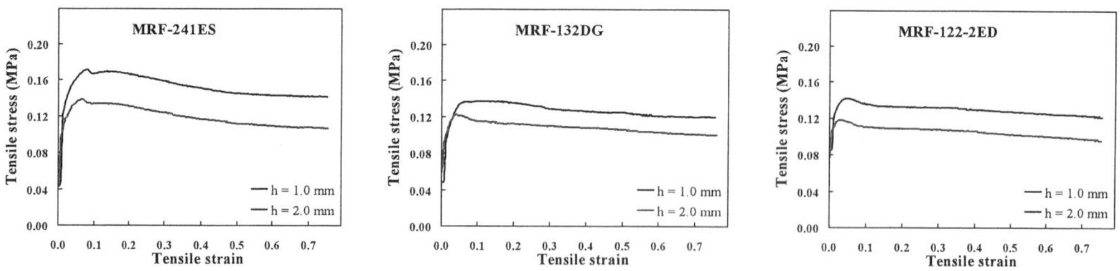

(a) (b) (c)

Fig. 4. Effect of the initial gap size under a constant tensile speed of 1.5 mm/min and a constant applied current of 1.2 Amps for (a) MRF-241ES, (b) MRF-132DG and (c) MRF-122-2ED.

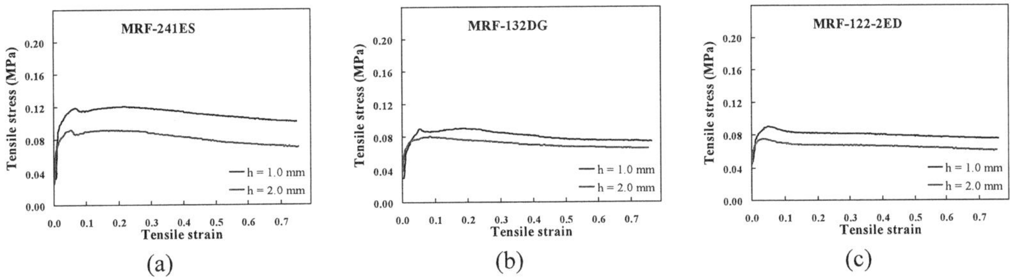

Fig. 5. Effect of the initial gap size under a constant tensile speed of 1.5 mm/min and a constant applied current of 0.8 Amps for (a) MRF-241ES, (b) MRF-132DG and (c) MRF-122-2ED.

At smaller initial gap size (1.0 mm), higher values of tensile stresses were found when a constant current of 1.2 Amps was used. More magnetic flux lines penetrated the MR fluids as the distance between the upper and lower cylinders became smaller. These results were expected since the applied magnetic field was the strongest at smaller initial gap sizes. However, the effect of initial gap size factor under tension loading disagreed with its effect under compression loading tests [4]. The tensile stress was dependent on the structure strengthening effect due to particle chains restructuring and the decrease of the magnetic field as the gap size increased. Therefore, the tensile stress can be considered proportional to the magnetic field strength.

c. Effect of tensile speed

Stress-strain relationships under different tensile speeds are shown in Figs. 6 and 7. The tests were conducted at 1.0 and 2.0 mm of initial gap sizes. The current passing through the coil was 1.2 Amps for all cases. It could be observed that the stress-strain curves were well overlapped for all values of the tensile speed.

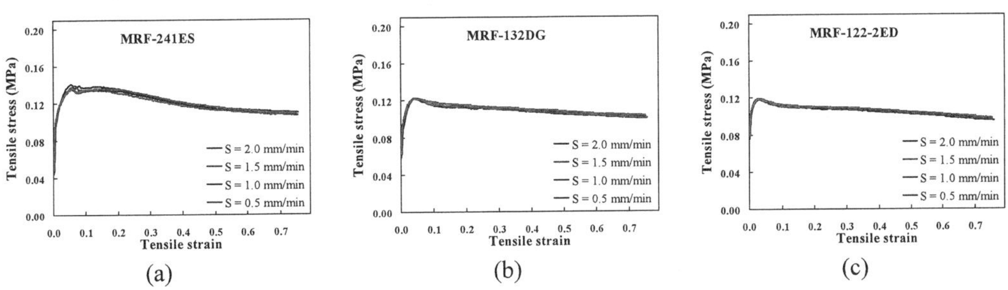

Fig. 6. Effect of the tensile speed under a constant current of 1.2 Amps at initial gap size of 2.0 mm for (a) MRF-241ES (b) MRF-132DG and (c) MRF-122-2ED.

The tensile speeds used in this study were in the range of 0.0083 to 0.0333 mm/s, which were comparable to speeds used by other researchers. For instance, Tian et al. [7] used a constant speed of 0.0524 mm/s to investigate the tensile behaviour of ER fluids. It might be anticipated that the effect of the tensile speed was not significant on the tensile stress since the structure formation occurred on a time scale of milliseconds [10,12]. Even though, in this study, various values of the tensile speeds were applied to move the upper cylinder, the speeds were still considered very slow compared to the timescale of structure formation. Based on the speed and initial gap size ranges experimented in this study, the tension process time ranges were, approximately, between 30 and 240 seconds. Much higher speed values are necessary to detect the stress variance with speed, on a millisecond timescale. However, the use of high speeds was prohibited by the gap size restriction in the design and the motor movement mechanisms in the UTS machine.

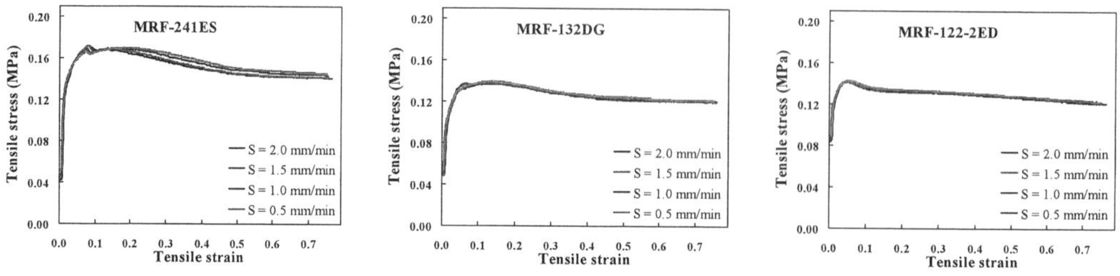

(a) (b) (c)

Fig. 7. Effect of the tensile speed under a constant current of 1.2 Amps at initial gap size of 1.0 mm for (a) MRF-241ES (b) MRF-132DG and (c) MRF-122-2ED.

Conclusion

Effects of the applied current, the initial gap size and the tensile speed on three types of MR fluids in tension mode have been investigated. Experimental results were presented starting with different electrical currents to generate magnetic fields, using different initial gap sizes and tensile speeds. The results revealed that the stress-strain curves of MR fluids can be divided into three different regions, due to changes in the MR fluid characteristics. The tensile stress was assumed to be dependent on the structure strengthening effect due to particle chains restructuring and decreasing magnetic field as the gap size increased. The results emphasized that high values of stress occurred when the electrical current was high. Furthermore, the tensile stress-strain curves were dependent on the initial gap size. Curves showing the relationships between the stress and the strain were constructed for two different initial gap sizes. At smaller initial gap sizes, the values of stress were larger. However, contradicting with the effects of the applied current and the initial gap size, the tensile speed had a small effect on the stress-strain relationship of MR fluids. The effect of tensile speed can be further studied in the future using higher tensile speeds for the loading process. The latter recommendation requires a new test rig design to detect the comparatively faster effects of high tensile speeds.

References

[1] K. Kim and D. Jeon: J. Intell. Mater. Syst. Struct. Vol. 10 (1999), p. 779.

[2] M.R. Jolly, J.W. Bender and J.D. Carlson: J. Intell. Mater. Syst. Struct. Vol. 10 (1999), p. 5.

[3] S. Sassi, K. Cherif, L. Mezghani, M. Thomas and A. Kotrane: Smart Mater. Struct. Vol. 14 (2005), p. 811.

[4] S.A. Mazlan, N.B. Ekreem and A.G. Olabi: Smart Mater. Struct. Vol. 16 (2007), p. 1678.

[5] S.A. Mazlan, N.B. Ekreem and A.G. Olabi: J. Mater. Process. Tech. Vol. 241 (2008), p. 780.

[6] S.A. Mazlan, N.B. Ekreem and A.G. Olabi: J. Phys. D: Appl. Phys. Vol. 41 (2008), 095002.

[7] Y. Tian, Q. Zou, Y. Meng and S. Wen: J. Appl. Phys. Vol. 94 (2003), p. 6939.

[8] Y. Tian and Q. Zou: Appl. Phys. Lett. Vol. 82 (2003), p. 4836.

[9] S.B. Choi, H.J. Choi, Y.T. Choi and N.M. Wereley: J. Appl. Polym. Sci. Vol. 96 (2005), p. 1924.

[10] J.D. Carlson and M.R. Jolly: Mechatronics Vol. 10 (2000), p. 555.

[11] S.A Mazlan, A. Issa and A.G. Olabi: Adv. Mater. Research Vols. 47-50 (2008), p. 242.

[12] K. Kim and D. Jeon: J. Intell. Mater. Syst. Struct. Vol. 10 (1999), p. 779.

Solid State Phenomena Vol. 154 (2009) pp 133-138
online at http://www.scientific.net

Texture analysis of hot rolled Ni-Mn-Ga alloys

H. Morawiec[a], T. Goryczka, A. Drdzeń, J. Lelątko, K. Prusik

University of Silesia, Institute of Materials Science, Bankowa 12, 40-007 Katowice, Poland

[a]email: morawiec @us.edu.pl

Keywords: Ni-Mn-Ga, hot-rolling, texture, EBSD

Abstract. The studies were carried out on three Ni-Mn-Ga polycrystalline alloys with the 10M martensite and the $L2_1$ parent phase. The specimens were cut along the [001] and [110] directions of the columnar grains from the ingots of either round or flat shape. The samples were heated up to 1000^0C and rolled in one direction in several steps. The final reduction of the samples thickness was: 28%, 36%, 57% and 69%. The texture of the initial [001] orientation rolled with the highest deformation 60% shows the following orientation components: {111}<112>, {112}<011>, {111}<011> which are typical for deformed and recrystallized metals of the A2 structure. The texture of the specimen with the initial [110] orientation rolled with 69% can be described as a fibre texture scattered around the <110> direction.

Introduction

The main theoretical principles for selection ferromagnetic shape memory alloys based on Heusler compounds are generally known and experimentally proved for single crystals of Ni-Mn-Ga alloys [1, 2]. Extension of these results to polycrystalline materials would markedly increase the application capabilities of this class of materials. It is also understood that magnetic field induced strain (MFIS) of polycrystalline alloys is remarkably smaller than that of single crystals because of the randomly oriented grains and twin variants in martensite. One of important factor, which controls the MFIS effect is the magneto-crystalline anisotropy of the martensite with $c/a<1$.

A crystallographic texture of the polycrystalline alloy can significantly improve this effect because the preferred orientation is the <001> direction which is the easy magnetization direction. The studies to obtain textured polycrystalline Ni-Mn-Ga alloys encompass the directional solidification [3], melt spun ribbon processing [4] and directionally deformation [5, 6].

The plastic deformation this alloys done by compression at high temperature was studied by Böhm et al. [3]. They stated the essential lowering the flow stress at 1000 °C compared to 800 °C. The microstructure of the martensite after deformation shows the presence of twins. The possibilities of deformation to be done by hot rolling the small ingots of Ni-Mn-Ga alloys placed in iron cylinder have been shown by Bessegnini et al. [5]. It was stated, that the deformation has a beneficial effects such as homogeneity of the alloys and possibility to close pores and holes in the as cast material.

More detailed studies of the deformation by hot rolling were published by Bin et al. [6]. The authors, basing on comparing the diffraction patterns before and after hot rolling stated increase of diffraction peaks indexed as (202), (400) and (323) and concluded the presence of the (400) texture.

The aim of this work was to study the texture after hot rolling the Ni-Mn-Ga alloys with different deformation degree and different initial orientation.

Experimental procedure

The studied polycrystalline alloys were prepared by melting in an induction furnace under argon atmosphere using 99.99% pure elements. The chemical composition determined by EDS method, structure at room temperature and initial crystallographic orientation of the specimens are listed in Table 1. The specimens in rectangular shape were cut along the [110] and [001] direction of the columnar grains from the flat or round shape ingots of alloys 1 and 2. The relation between initial crystallographic orientation of the samples and the rolling direction is shown in Fig. 1. The alloy 3 was in the parent phase state and the samples were cut from a large flat ingot with random oriented grains.

[010] – fiber axis [110] – fiber axis

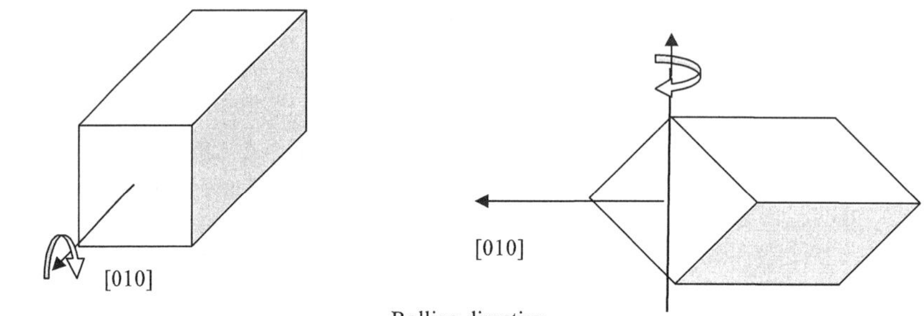

Rolling direction Rolling direction

Figure 1. Initial crystallographic orientation related to the rolling direction.

The rolling was carried out on the samples imbedded in a steel channel bar and „sealed" in a flat pipe. The samples were heated up to 1000°C and rolled in one direction in several steps. Before each step the sample was reheated. The final reduction of the samples thickness was 28%, 36%, 57% and 69%. The texture was studied by EBSD and X-ray diffraction using the {001}, {110} and {111} pole figures. The Kikuchi patterns are generated within a 50 nm deep layer, so a sophisticated surface preparation is needed. A final polishing step using colloidal diamond suspension was used for this purpose.

Table 1. Chemical composition, structure and initial orientation of the alloys.

Alloy	Ni [at%]	Mn [at%]	Ga [at%]	M_s [K]	Structure at room temp.	Initial orientation
1	49,13	27,48	23,39	328	10M	<110>
2	51,5	28,8	19,8	366	10M	<010>
3	51,3	27,6	20,1	290	$L2_1$	random orientation

The EBSD measurements were carried out using electron scanning microscope (JEOL 6480) equipped with Nordlys detector supported by computer with HKL Channel 5 system. In order to collect grain orientation from large surface as possible the experiments were done at magnification of 100x and grid step size of 4μm,. Every orientation map consisted of 4 mapping jobs.

In the case of X-ray diffraction analysis the Philips PW 1130 diffractometer was used with texture goniometer. The X'Pert Pro diffractometer was applied for phase identification and precise lattice determination.

Results and discussion

First it was stated that at room temperature alloys 1 and 2 are in martensitic phase. The structure of the martensite was identified by X-ray diffraction method as 10M. The TEM studies have shown the five layered structure. For alloy 1 it is shown in Fig.2. The diffraction pattern of the annealed at 900°C specimen shows the presence of additional four spots. The initial orientation of the as cast specimen can be seen in the pole figures based on the EBSD scattered data in Fig. 3. The pole figures show the presence of a large scattered [110] fiber texture. To determine the texture in the tetragonal martensite with c/a=0.94 it was assumed that with a sufficient accuracy (1,5°) the standard projection for regular crystal system may be used. For this reason the texture axis in Table 2 includes the angle ρ which is formed between the fiber axis and normal to the {hkl} plane for a regular system. The inverse pole figure in Fig. 3b presents the <001> normal to the one of the surfaces of the sample – confirming the <110> fiber axis.

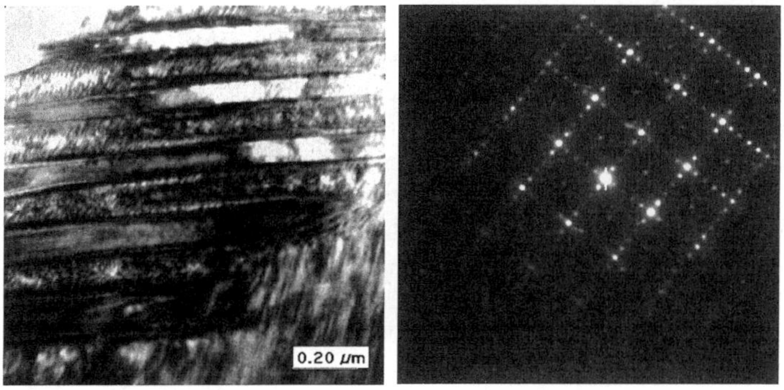

Figure 2. Morphology of martensite and diffraction pattern of 5 layered structure in alloy 1 annealed at 900°C.

Table 2 Angle between the texture axis and normal of the diffraction plane

Diffraction plane	ρ° <110>	ρ° <112>
{001}	45; 90	35; 66
{110}	60; 90	30; 55; 73; 90
{111}	35; 90	19; 62; 90

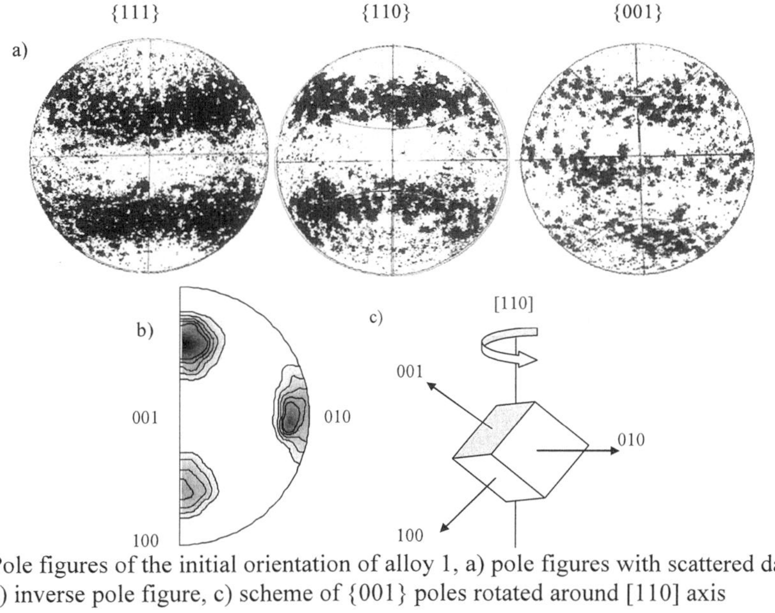

Figure 3. Pole figures of the initial orientation of alloy 1, a) pole figures with scattered data, b) inverse pole figure, c) scheme of {001} poles rotated around [110] axis

Computer treatment of the scattered data allows to contouring the pole figure which selects the preferred orientation and reveals their distribution.

The microscopic observation of the hot rolled specimens has shown a typical recrystallized microstructure. Thus the deformed at 1000^{0}C samples cannot be treated as pose deformation textures but should be related to recrytstallization. However, this texture is changing with the amount of total deformation degree as can be seen on the pole figures in Fig. 4a and 4b. The pole figures in Fig. 4a may be approximately described with a 5-10 degree deviation of the projections of the following texture sheet components: {111}<112>, {111}<110> and {110}<112>. Figure 4b depicts the pole figures for the alloy 2 which undergoes the total deformation by 69%. These pole figures show the overlapping of the typical orientation components: {111}<112>, {111}<110>, {112}<110> and {110}<112> and for this reason on the pole figures 4b have been drawn the circles related to the fiber axis for <110> and <211> directions which embrace the poles. Thus, approximately the texture for the sample with highest deformation maybe described by fiber axis. The confirmation of this is given by the inverse pole figures in Fig. 4c, which show both of the poles <110> and <211> on the surface of the specimen.

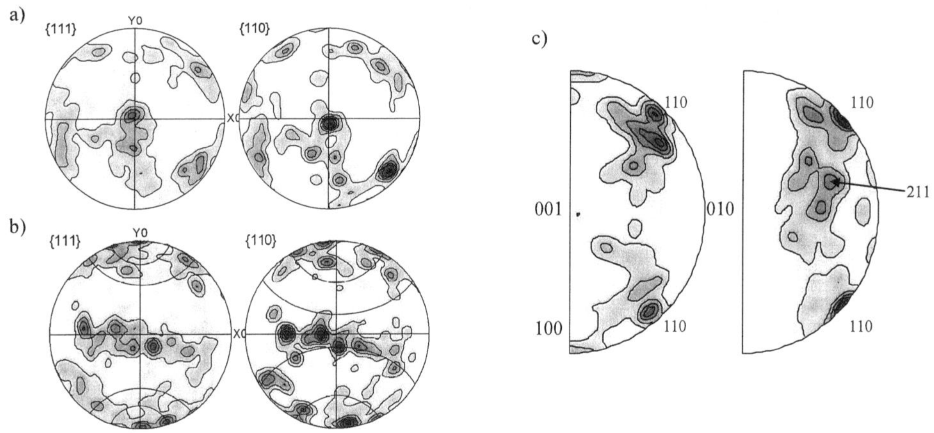

Figure 4. EBSD (111) and (110) pole figures for the deformed samples of alloy 1, pole figures after 36% deformation (a) and after 69% deformation (b), inverse figures for sample deformed by 69% (c).

The alloy 2 has also the five – layer structure martensite, but sample in the as cast state were cut along [001] direction. Fig. 5a shows the EBSD pole figures of the as cast state. Sample annealing at 900°C for one hour doesn't destroy the [010] fiber texture. As can be concluded from the (001) and (110) pole figures for both states of the specimen, beside the fiber texture, the sheet component (100) <010> can be also distinguished. Microscopic observation shows a presence of the elongated grains in the sample before the deformation.

The alloy 3 possesses at room temperature the $L2_1$ parent phase structure and randomly oriented small grains. The texture of this alloy deformed by 59% is shown in Fig. 6. The EBSD pole figures shows a strong sheet component {110}<001> and additional the {111}<110> but on the (001) pole figure the <110> fibre texture is present. In order to increase the plasticity of the Heusler compounds with $L2_1$ supperstructure the rolling was carried out in the B2 range - at 1000^{0}C. Thus, the expected texture is similar to that obtained for metals with the A2 (B2) type of structure. The sheet with random grain orientation exhibits a recrystallization texture described with the following components of ideal orientations: {110}<001> and {111}<110>. The superposition of these sheet components empowers to describe it using the fibre texture obtained by rotating the privileged orientations if a definite angle around the axis like <110> = ND (normal direction). The pole figures shown in Fig. 4b and 7 may be interpreted as an axial texture with large scattering of the poles. Some additional clusters around the <120> and <122> directions can be also recognized.

In order to compare the result obtained by the EBSD method, additional texture studies were carried out by X-ray diffraction method. Fig. 7 shows the (010) pole figures for the deformed alloys: alloy 1 by 69% alloy 2 by 60% and alloy 3 by 59%. Pole figures for alloy 1 and 2 can be described approximately by the following sheet components: (111) <011>, (111) <112>, and (112) <011>. The pole figure for alloy 3 shows the effectiveness of deformation by hot rolling to form the preferred orientation. As it is shown on the pole figures (Fig. 7) obtained by X-ray technique the circles of the <110> and <211> fiber texture has been drawn, which coincide with some poles of the sheet components. To achieve a higher quality of the X-ray pole figures a larger size specimen is required. Generally one can conclude that both of the compared texture studying techniques has given comparable results.

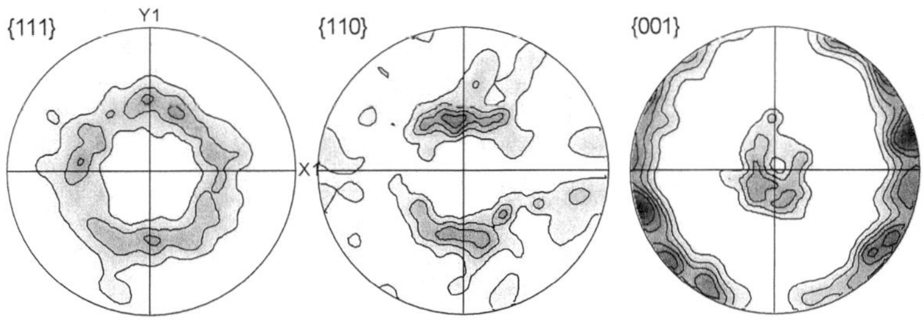

Figure 5. EBSD pole figures for alloy 2 with the initial fiber texture [001]

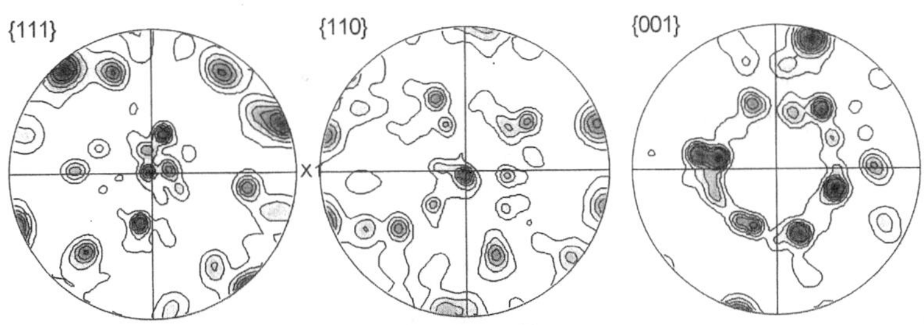

Figure 6. EBSD pole figures of deformed by 59% alloy 3.

In order to increase the plasticity of the Heusler compound with $L2_1$ superstructure the rolling was carried out in B2 type structure at relatively high temperature of 1000^0C. Thus, expected texture should be similar to those obtained for metals and alloys with structure of the A2 (B2) type.

The sheet with random grain orientation exhibits a recrystlization texture described with the following components of ideal orientations: {110}<001> and {111}<110>. The superposition of these sheet components of the texture empowers to describe it using the fibre texture obtained by rotating the privileged orientations of a definite angle around the axis like <110>=ND (normal direction). The pole figures shown in Fig. 4, 6 and 7 may be interpreted as axial texture with large scattering of the poles. Some additional clusters around the <120> and <122> directions can be also recognized.

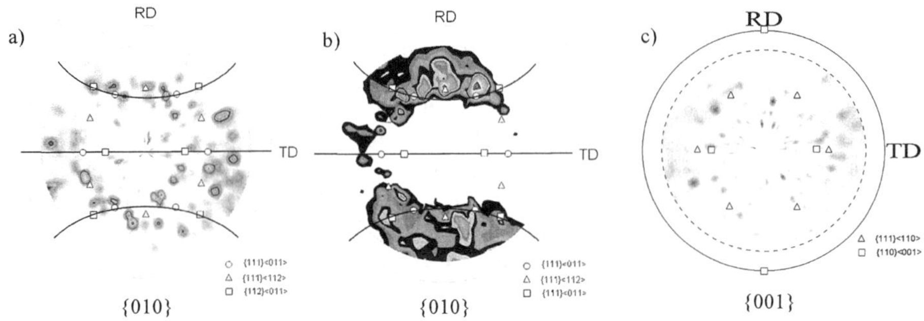

Figure 7. X-ray pole figures for alloy 1 (a) alloy 2 (b) and alloy 3 (c).

Conclusions

- The recrystallization texture of the hot rolled sheets of Ni-Mn-Ga alloys shows the following sheet texture components: (111)<112>, (111)<110>, {110}<001> and (112)<011> .
- The superposition of these components allows to describe approximately grain orientation as a fibre texture with the <110> and <211> axis.
- The alloy with random grain distribution in initial state reveals after rolling presence of the most favorable texture component: {110}<001>.
- Hot rolling at 1000^0C is a good method for plastic deformation of Ni-Mn-Ga.
- EBSD method had been successfully applied to examine texture of large grain ferromagnetic Ni-Mn-Ga shape memory alloys.

Acknowledgments: This work was supported financially by the Ministry of Science and High Education (project PBZ-KBN-115/T08/2006).

References

[1] A. A.Likhachev, K. Ullakko: Lett. A275 (2000), p. 142

[2] A. Sozinov, A.A. Likhachev,N. Lanska, K. Ullakko: Appl. Phys. Lett. 80 (2002), p. 1746

[3] A. Böhm, S. Roth, G. Nauman, W.G. Drossel, R. Negebauer :
Mater. Sci. Eng. A 481-482 (2008), p. 266

[4] Z.H. Liu, X. X. Zhang, M. Zhang, X. F. Dai, H. N. Hu, J. L. Chen, G. H. Wu:
Phys. Letters A 329 (2004), p. 214

[5] S. Besseghini, E. Villa, F. Passaretti, M. Pini, F Bonfanti: Mat. Sci. Eng. A 378 (2004), p. 415

[6] L. Bin, W. Hua-Bin, Liu Yan, L. Ji-Xuan, W. Hong Lei:
Trans. Nonferrous Met Soc. China 16 (2006), p. 843

Solid State Phenomena Vol. 154 (2009) pp 139-144
online at http://www.scientific.net

Monte Carlo Study of Magnetostructural Phase Transitions in $Ni_{50}Mn_{25+x}Sb_{25-x}$ Heusler Alloys

V.V. Sokolovskiy[1, a], V.D. Buchelnikov[1,b], S.V. Taskaev[1,c]

[1] Chelyabinsk State University, 454021 Chelyabinsk, Russia

[a]vsokolovsky84@mail.ru, [b]buche@csu.ru, [c]tsv@csu.ru

Keywords: shape – memory Heusler alloys, structural and magnetic phase transitions.

Abstract. On the basis of Monte Carlo simulations, we investigate the temperature dependence of magnetization of Ni-Mn-Sb Heusler alloys, in which the part of the Mn atoms interact antiferromagnetically. It is shown that this antiferromagnetic exchange is responsible for existence of several magnetic phase transitions in the Heusler alloys. For a certain set of parameters of the model Hamiltonian we obtain coupled martensitic and magnetic phase transitions. Results of simulations agree in an excellent way with the experimental magnetization curves.

Introduction

In recent years, Ni-Mn-X (X =In, Sn, Sb) alloys has been attracting attention in view of their unique properties, such as the shape memory effect, the magnetocaloric effect, the magnetoresistence and other properties are associated with the martensitic transition (MT) [1, 2]. These properties are applicable in developing actuator materials and materials for magnetic refrigeration.

Recent experiments have shown that in Heusler Ni-Mn-X (X = In, Sn, Sb) alloys, the structural phase transition from the paramagnetic (PM) high-temperature cubic (austenitic) phase to the ferromagnetic (FM) low-temperature tetragonal (martensitic) phase can occur [3]. It is well known, that in stoichiometric $Ni_{50}Mn_{25}X_{25}$ alloys all Mn atoms on its regular sublattice site interact FM. However, recent experiments on non-stoichiometric $Ni_{50}Mn_{25+x}X_{25-x}$ alloys revealed that in the martensite, the excess of Mn atoms occupying the sites of X atoms, interact with Mn atoms on the Mn sublattice sites antiferromagnetically (AF) [4]. Furthermore, the AF interaction is enhanced due to the MT, which is also proved by the drop in thermomagnetization curves [4, 5]. Thus, in the martensitic phase of $Ni_{50}Mn_{25+x}X_{25-x}$ alloys the FM and the AFM interactions coexist. Therefore, in these compounds the phase transition from the mixed AFM – FM martensite to the FM austenite and from FM austenite to PM austenite with increasing temperature is experimentally observed [3 - 5].

From a theoretical point of view, it is a demanding task to try to obtain all phases in self-consistent way. The *ab initio* calculations for the austenitic and martensitic phase and the resulting magnetic exchange constants of Ni_2MnGa, Ni_2MnIn and non- stoichiometric $Ni_{50}Mn_{34}In_{16}$ and $Ni_{50}Mn_{36.25}In_{13.75}$, using KKR-CPA method [6] clearly show the competition of FM and AF interactions. For example, for $Ni_{50}Mn_{34}In_{16}$ the AF exchange constant is -6.0 meV, for Ni-Mn-Sn the values of magnetic exchange constants (5.0 meV for the FM interaction and -5.0 meV for the AF interaction) for nearest neighbors, which was obtained from *ab initio* calculations [7]. Performing Monte Carlo simulations for the underlying Heisenberg model using the *ab initio* magnetic constants [6] shows however that it is difficult to obtain the sequence of structural and magnetic 1st-order-like phase transitions as observed in experiment [7].

A more refined Heisenberg-like model, which has been used for the description of magnetocaloric properties of rare-earth based and Heusler compounds, was introduced in Ref. [8, 9]. In addition to the model, which allows only a magnetic transition of 2nd-order without taking into account the MT, authors of [10] used a model for the description of the pre-martensitic phase transformation in Ni_2MnGa. For the description of the magnetic and structural part, the authors used the Ising model and the degenerated Blume-Emery-Griffiths (DBEG) model [10], respectively.

Another question is whether the physical properties of Ni-Mn-X would require considering also the softening of phonons associated with pre-martensitic or MT observed in some of the alloys close to stoichiometric composition [11]. In the most pronounced cases, the phonon softening involves energy changes of the order of 1-2 meV. In DBEG model [10], authors considered that the structural part the exchange constant is proportional to the soft phonon energy at the MT. In our Monte Carlo simulations of magnetic properties of Ni-Mn-Sb alloys using an extended BEG model, which allows for coupled magnetostructural phase transitions from the mixed AF – FM martensite to the FM austenite and the magnetic phase transition from FM austenite to PM austenite.

The Theoretical Model

In presence of the MT, the martensitic phase may exhibit several low-temperature variants. These variants can be described by lattice distortions (compression or expansion) during the MT along the *x, y, z* axes, which for the cubic phase yields six structural variants. The austenitic phase corresponds to a phase with zero displacement, i.e. $x = y = z = 0$. Thus, during cooling, the austenite may choose any of the six variants, and can be thought of as having a high degeneracy. In our model we consider two variants of martensite with the lattice deformation along $\pm x$ or $\pm y$ or $\pm z$ axes.

In the model we use a cubic lattice with periodic boundary conditions. Since the magnetic moments of the Ni atoms are small compared to the Mn moments and since the X atoms not have spin moment, we omit the contributions from the Ni and X atoms. Thus, in our model all lattice sites are occupied by Mn atoms. For $Ni_2Mn_{1+x}X_{1-x}$ alloys model we assume that a certain part of the Mn atoms on the lattice interact AF with each other and others Mn atoms. The initial configuration of these Mn atoms is randomly chosen and was determined from experimental compositions.

The system can be representing as two interacting magnetic and structural parts. For the magnetic part, we use the *q-state* Potts model which allows simulating the 1st order magnetic phase transition from FM to PM phases [12]. Here q is number of spin states. In our case we consider q=5 spin states because in Ni-Mn-X the spin state of Mn corresponds to S=4/2 with $2S$+1 possible spin projections. Hence, we have $q = (1 ,.., 5)$, a five-states Potts model. The structural part is described by the DBEG model which allows a structural transition from the cubic phase to the tetragonal one [10].

The generalized Hamiltonian (Eq. 1) include of three parts: magnetic part (Eq. 2), elastic part (Eq. 3) and the magnetoelastic interaction (Eq. 4).

$$H^* = H^*_m + H^*_{el} + H^*_{\text{int}}, \tag{1}$$

$$H^*_m = -J^*_{fm} \sum_{<i,j>}^{NN} \delta_{S_i,S_j} - J^*_{afm} \frac{T^* - T^*_{afm}}{T^*_{afm}} \sum_{<i,j>}^{NN} \delta_{S_i,S_j} - h^* \sum_{i}^{N} \delta_{S_i,S_g}, \tag{2}$$

$$H^*_{el} = -\sum_{<i,j>}^{NN} \sigma_i \sigma_j - K^* \sum_{<i,j>}^{NN} (1-\sigma_i^2)(1-\sigma_j^2) - T^* \ln(p)(1-\sigma_i^2), \tag{3}$$

$$H^*_{\text{int}} = 2U^* \sum_{<i,j>}^{NN} \delta_{S_i,S_j} (\frac{1}{2} - \sigma_i^2)(\frac{1}{2} - \sigma_j^2) - \frac{1}{2} U^* \sum_{<i,j>}^{NN} \delta_{S_i,S_j}, \tag{4}$$

$$H^* = H/J, J^*_{fm} = J_{fm}/J, K^* = K/J, U^* = U/J, T^* = k_B T/J, T^*_{afm} = k_B T_{afm}/J, h^* = g\mu_B H_{ext}/J. \tag{5}$$

Here, J_{fm} and J_{afm} are the FM and AF exchange constants; J, K are exchange constants in the structural subsystem; T is the actual temperature; T_{afm} is the temperature at which the AF interaction change its sign, see second term in (Eq. 2); $\delta S_i,S_j$ is the Kronecker symbol which limits the spin-spin interaction to interactions between the same q-states only; S_g is kind of a ghost-spin, whose direction is that of the external magnetic field, a spin parallel to the ghost spin is favored by positive H_{ext}; g is the Lande factor; μ_B is the Bohr's magneton; k_B is the Boltzmann constant; U is the constant of magnetoelastic interaction and p is the degeneracy factor. The variable σ_i = 1, 0, -1 represents the deformation state on each sites on the lattice, where $\sigma_i = 0$ denotes the undistorted phase and states $\sigma_i = \pm 1$ the distorted phases. Sums are performed over all nearest-neighbor pairs.

The AF interaction term in (Eq. 2) limits the AF interactions to the low-temperature phase. For $T > T_{afm}$ this interaction term will change sign resulting in additional FM interactions. Accordingly, above this temperature, enhance FM between all Mn atoms occur. Since, this means that there will be no transition from FM austenite to PM austenite, we allow the the maximal value of the factor $(T^*-T^*_{afm})/T^*_{afm}$ to be unity, i.e. for temperatures $T^*>2T_{afm}^*$ the prefactor of the second term (Eq. 2) is assumed not depend on temperature any more and is set equal to $-J_{afm}^*$. By the help of this an admission we can describe several magnetic transitions from the PM phase to different magneto-ordered phases with decreasing temperature like the FM, AF phases or mixed FM-AF phases.

In the elastic part (Eq. 3), the first (second) term characterizes the interaction between single strains σ_i in the martensite (austenite), respectively. Large values of K^* will stabilize the *pure cubic* phase in which $\sigma_i = 0$ holds. Moreover, the parameter K^* determine and control the order of the transition. In case of low values of K^*, the cubic phase will be stabilized in the high temperature region where we find an equal population of the strain variables $\sigma_i = 0$ and $\sigma_i = \pm 1$, this case corresponds to an *average cubic* phase [10]. The last term in (Eq. 3) characterizes a temperature-dependent degeneracy factor p for the cubic phase [10].

In the *q-state* Potts model the magnetization of the system is presented by (Eq. 6).

$$m = \frac{1}{N}\frac{qN_{\max} - L^3}{q-1}, \tag{6}$$

where N_{max} is the maximal number of same magnetic states in sites of the lattice.

As in DBEG model [10] the strain order parameter is calculated by (Eq. 7).

$$\varepsilon = \frac{1}{N}\sum_{i}^{N}\sigma_i , \tag{7}$$

For $\varepsilon = 0$ in the DBEG model we have two cubic states (*pure cubic* or *average cubic*) depending on the values of strain variables σ_i. For $\varepsilon = 1$, we find the marteniste for one of variants σ_i=1 or σ_i= −1.

Numerical Results

Following we discuss numerical results for $Ni_{50}Mn_{25+x}Sb_{25-x}$ alloys obtained by the Monte Carlo method. The corresponding simulations have been carried out using a standard Metropolis algorithm [13]. The changes in the spin states q and σ_i are treated independently and are accepted or rejected according to the single-site transition probability $W=\min\{1, e^{-\Delta H^*/T^*}\}$. For simulations of the cubic lattice, we used periodic border conditions and 6 near-neighbors for each site. The number of sites in the lattice was equal to $N=15^3$. The time unit is one Monte Carlo step, which consists of N attempts to change the q and σ_i variables. For given the temperature the number of Monte Carlo steps on each site is 10^5. We start the Monte Carlo simulations from the FM martensite with q=1 and σ_i =1. The various quantities are averaged over 400 configurations taken every 100 Monte Carlo steps and discarding the first 10^4 Monte Carlo steps for equilibrium. The degeneracy factor and the Lande factor were taken as p=2, g=2. In order to find the spin state at each lattice site, we choose a random number $0 < r < 1$ and fix the value of q according to the scheme: if $0 \le r \le 1/5$, then $q = 1$ if $1/5 < r \le 2/5$, then $q = 2$; if $2/5 < r \le 3/5$, then $q = 3$; if $3/5 < r \le 4/5$, then $q = 4$; if $4/5 < r \le 5/5$, then $q = 5$.

In the Monte Carlo simulations we adopted the parameter values listed in Table I in order to describe the magnetic and structural properties of the Ni-Mn-X alloys in an optimal way.

Table 1. Model parameters for $Ni_{50}Mn_{25+x}Sb_{25-x}$ alloys.

	J_{fm} [meV]	J_{afm} [meV]	U [meV]	K [meV]	T_{afm} [meV]
x=7	4.6	4.6	-0.4	0.0	0.80
x=12.5	5.6	4.4	-1	-0.2	1.42
x=13	6.1	4.6	-2	-0.08	1.36
x=13.5	6.2	3.46	-2.6	-0.02	1.41

The information about the value of the structural exchange interaction J was taken from experimental data of the phonon dispersion curves of the Ni-Mn-Ga. As it known from [11] that the exchange constant J is proportional to the energy of the soft phonon $\hbar\omega$, and for Ni-Mn-Ga value of J at the MT is equal 2 meV approximately. Values of our FM and AF exchange constants J_{fm} and J_{afm} (see Table I) nearly coincide with values of the magnetic constants for Ni-Mn-X obtained from *ab initio* calculations [7]. In order to determine values of parameters K, which characterizes the MT, and values of U, we used following conditions: U<0 and K<0 [10]. It is necessary to note that for values of K>0 by the help of the BEG model it is possible to obtain the martensitic, pre-martensitic and magnetic phases transitions at temperatures $T_m<T_I<T_C$ [10]. Since the pre-martensitic transition in Ni-Mn-X has not been observed experimentally, so we consider that values of $K < 0$. In our model for determining the distance between T_m and T_C we used different values of U, K and T_{afm}.

The results of the simulations are presented in Figs. 1-3. Fig. 1 and 2 show the temperatures dependences of the magnetizations and strain order parameters (tetragonal distortion ε) for $Ni_{50}Mn_{25+x}Sb_{25-x}$ alloys. Here, the filled circles and solid lines are results of our Monte Carlo simulations and the open circles are experimental results which have been taken from [4].

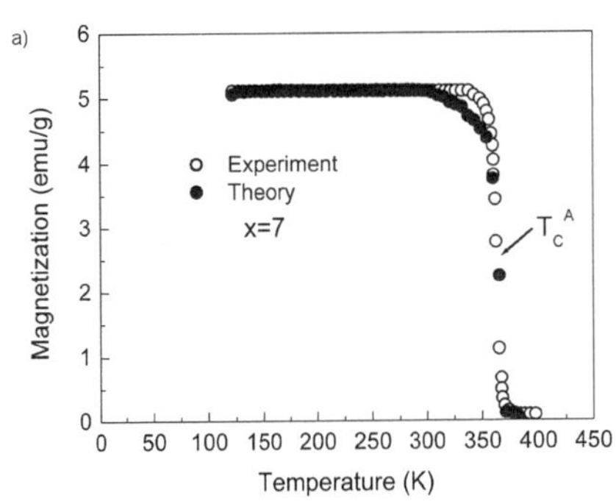

Fig.1a. Temperature dependences of the magnetization of $Ni_{50}Mn_{32}Sb_{18}$ alloy in a magnetic field 0.01T.

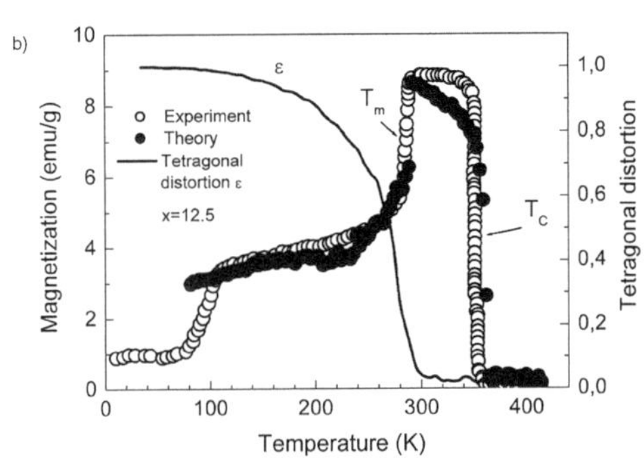

Fig.1b. Thermomagnetization curves and tetragonal distortion of $Ni_{50}Mn_{37.5}Sb_{12.5}$ alloy is absent of a magnetic field.

Fig. 1a shows the temperature dependences of the magnetizations for the $Ni_{50}Mn_{32}Sb_{18}$ in a magnetic field 0.01 T. For this composition the structural transformation from the cubic state to the tetragonal one is absences and the $Ni_{50}Mn_{32}Sb_{18}$ alloy has the austenitic phase in whole a temperature region [4, 5]. Therefore, in our calculations we have taken value of parameter K, which determines the temperature of the structural transition, as equal 0 (Table 1). So, in Fig. 1a, we show the magnetic transition from the FM austenite to PM austenite at 360 K.

Fig. 1b represented the magnetization and the strain as function of a temperature for $Ni_{50}Mn_{37.5}Sb_{12.5}$ alloy is absent of a magnetic field. The two magnetic phase transitions are now at 350 K and at 280 K, respectively. We obtain from the simulations a PM-FM transition in cubic state at 350 K. The second transition at 280 K describes the FM to mixed FM-AF phase transition which is again accompanied by a tetragonal transformation.

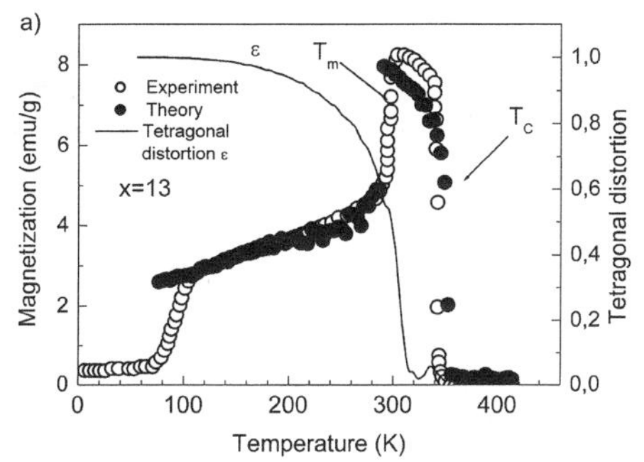

Fig.2a. Temperature dependences of the magnetization and tetragonal distortion of $Ni_{50}Mn_{38}Sb_{12}$ alloy is absent of a magnetic field.

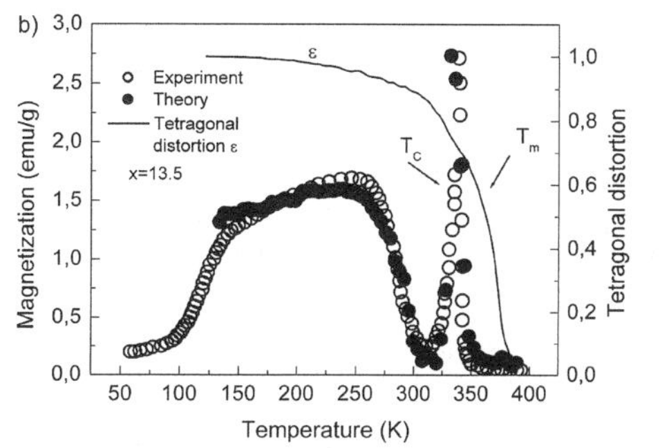

Fig.2b. Thermomagnetization curves and tetragonal distortion of $Ni_{50}Mn_{38.5}Sb_{11.5}$ alloy is absent of a magnetic field.

In Fig. 2a, we show theoretical and experimental results for $Ni_{50}Mn_{38}Sb_{12}$ alloy is absent of a magnetic field. We observe here two phase transitions, at 350 K and 300 K. At 350 K, we find the PM-FM transition in the austenite. The second transition is the coupled magnetostructural transition from the FM austenitic state to the mixed FM-AF martensitic state.

Figure 2b shows thermomagnetization curves and tetragonal distortion of $Ni_{50}Mn_{38.5}Sb_{11.5}$ alloy is absent of a magnetic field. As seen from figure, we may observe four phase transitions, at 360 K, at 340, at 320 K and at 270 K, respectively. The first transition at 360 K is the structural transition from the PM austenite to PM martensite; the second transition is the magnetic transition from the PM martensite to FM martensite; the third transition from FM martensite to PM or AF martensite, and the fourth transition from the PM or AF martensite to the mixed FM-AF martensitic state.

In order to test the predictability of our model, we have tried to describe the change of measured magnetization curves with composition for one system, only by allowing K to vary in order to obtain the experimental values for the structural transition, but leaving the other parameters unchanged. Values of model parameters: J_{fm}=5.1 meV, J_{afm}=4.5 meV, U=-1.8 meV, T_{afm}=1.34 meV. The parameter K was taken to be composition dependent: x=12, K=-0.6 meV; x=13.5, K=-3 meV; x=14, K=-6.2 meV; x=15, K=-31 meV; x=18, K=-150 meV. The resulting composition dependent theoretical and experimental magnetization curves are shown in Fig.3a and 3b. A comparison of theoretical and experimental results shows that the Hamiltonian (Eq. 1) allows to reproduces qualitatively the experimental trends [5].

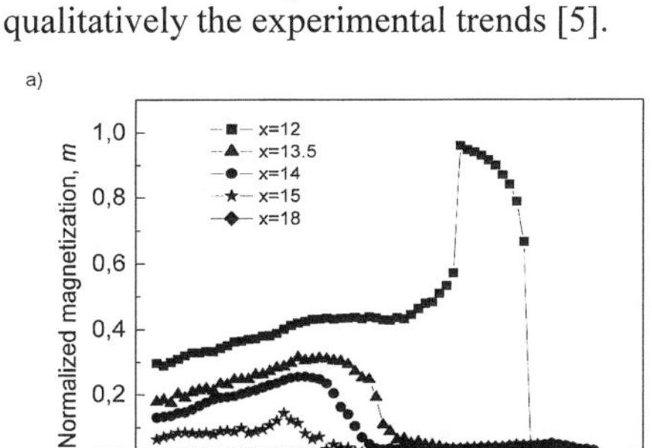

Fig.3a. Theoretical variation of the magnetization as function of temperature of $Ni_{50}Mn_{25+x}Sb_{25-x}$ for different compositions in a magnetic field 0.1 T as obtained from the Monte Carlo simulations.

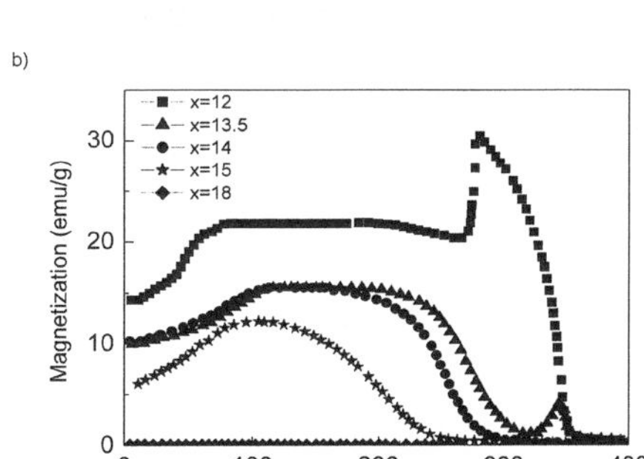

Fig.3b. Experimental variation of the magnetization as function of temperature of $Ni_{50}Mn_{25+x}Sb_{25-x}$ for different compositions in a magnetic field 0.1 T, data from [5].

Summary

In this work we have investigated the temperature dependences of the magnetization and strain of Heusler $Ni_{50}Mn_{25+x}Sb_{25-x}$ alloys using Monte Carlo simulations. Part of Mn atoms on simple lattice are assumed to interact only AF, whereby the concentration of these Mn atoms is determined by the experimental compositions. In addition to the magnetic interactions, the model Hamiltonian contains structural degrees of freedom allowing for a tetragonal deformation. The magnetic subsystem is described by *q-state* Potts model. For the structural subsystem we have used the DBEG model. The resulting model allows for a coupled 1st order magnetostructural phase transition from FM martensite to PM austenite. The AF interaction term in the model leads to several magnetic phase transitions from high temperature PM phase to different magnetoordered phases with decreasing of the temperature like the FM phase, the AF phase or the mixed FM - AF phase. The comparison of results of simulations and experiment is in some case satisfying.

Work was supported by grants RF and CRDF Y2-P-05-19, RFBR 05-08-50341, 06-02-16266, 07-02-96029-r-ural, 06-02-39030-NNSF, 05-02-19935-YaF_a RFBR and JSPS.

References

[1] P. Entel et al.: Mater. Sci. Forum Vol. 583 (2008), p. 21.

[2] R. Kainuma et al.: Appl. Phys. Lett. Vol. 88 (2006), p. 195513.

[3] Y. Sutou et al.: Appl. Phys. Lett. Vol. 85 (2004), p. 4358.

[4] M. Khan et al.: Appl. Phys. Lett. Vol. 91 (2007), p. 072510.

[5] M. Khan et al.: J. Phys. Condensed Matter Vol. 20 (2008), p. 235204.

[6] H. Akai and P.H. Dederichs: Phys. Rev. B Vol. 47 (1993), p. 8739.

[7] V.D. Buchelnikov et al.: Phys. Rev. B Vol. 78 (2008), p. 184427.

[8] E.P. Nobrega et al.: Phys. Rev. B Vol. 74 (2006), p. 144429.

[9] E.P. Nobrega et al.: Phys. Rev. B Vol. 72 (2005), p. 134426.

[10] T. Castan et al.: Phys. Rev. B Vol. 60 (1999), p. 7071.

[11] U. Stuhr et al.: Phys. Rev. B Vol. 56 (1997), p. 14360.

[12] F.Y. Wu: Reviews of Modern Physics Vol. 54 (1982), p. 235.

[13] D.P. Landau and K. Binder: *A guide to Monte Carlo simulations in statistical Physics* (Cambridge University Press, Cambridge, 2000).

Solid State Phenomena Vol. 154 (2009) pp 145-150
online at http://www.scientific.net

Microstructure and magnetic properties of two phase β+γ ferromagnetic Co-Ni-Al alloys

Wojciech Maziarz[1, a], Jan Dutkiewicz[1, b], Rafał Wróblewski[2,c], Marcin Leonowicz[2,d]

[1]Institute of Metallurgy and Materials Science, Polish Academy of Sciences, 25 Reymonta Str., 30-059 Krakow, Poland

[2]Warsaw University of Technology, Faculty of Materials Science and Engineering, Wołoska 141, 02-507 Warsaw, Poland

[a]nmmaziar@imim-pan.krakow.pl, [b]nmdutkie@imim-pan.krakow.pl,

[c]r.wroblewski@inmat.pw.edu.pl, [d]mkl@inmat.pw.edu.pl

Keywords: Co-Ni-Al ferromagnetic alloys, two phase structure, TEM, SEM structure, texture, magnetic properties

Abstract. The microstructure, texture and magnetic properties of two ferromagnetic alloys of composition $Co_{35}Ni_{37}Al_{28}$ and $Co_{37}Ni_{35}Al_{28}$ (in at. %) were investigated with optical microscopy (OM), analytical transmission electron microscopy (TEM), scanning electron microscopy (SEM) techniques as well as a vibrating sample magnetometer (VSM). The alloys were plastically deformed and heat treated in order to promote the martensitic transformation. Differential scanning calorimetry (DSC) revealed a drop of M_s temperature due to the increase of Co content in the alloys. The elongated twinned grains of size about 200 μm and small precipitates of γ phase were identified with the OM techniques. The TEM observations showed the twinned $L1_0$ non modulated martensite within the elongated grains and a small amount of ordered γ' phase after the heat treatment. The chemical composition of different martensite grains was established using point analyses of a HAADF-EDS technique. The magnetic phase transition temperatures were determined on the deformed samples using VSM method.

Introduction

The polycrystalline Co-Ni-Al alloys are considered the alternative to single crystalline Ni-Mn-Ga alloys for a magnetic shape memory application [1]. The maximum deformation which can be achieved in the Co-Ni-Al alloys in an external magnetic field is close to 6% and is lower than in Ni-Mn-Ga alloys (~10%) [2]. However, taking into account a lower cost of the constituent elements, higher plasticity and simplified technology of formation, these alloys have recently become interesting for future applications and have been investigated by several research groups [1].
The origin of plasticity of Co-Ni-Al alloys is due to the existence of a two phase region of beta (β) and gamma (γ) phases in the ternary phase diagram [3]. The γ (f.c.c.) phase is considered as a ductile one and at a certain composition of the alloy it can be plastically deformed at a relatively low temperature, because the (β+γ) phase region widens with decreasing temperature [4]. Finally, after hot deformation the alloys are annealed at temperature, at which the single β phase exists and cooled in order to promote the martensitic transformation. The martensitic transformation start temperature (M_s) can change in dependence on the composition of the alloys. Especially, an increase of aluminium content decreases the M_s temperature due to electron atom ratio (e/a) [5]. The lower content of aluminium in the alloys ensures better plasticity, but also a higher temperature stability of the β phase. Thus, it is very important to find a suitable balance between composition, structure and heat treatment of the Co-Ni-Al alloys. In the previous works [6-8] the results concerning the alloys with composition Co_{35+x}-Ni_{40-x}-Al_{25} (x = 0, 2.5, 5.0) were presented. It was shown that such a level of aluminium content assured satisfactory plasticity at 900°. However, the martensitic transformation effect was very low due to the occurrence of a significant amount of the γ phase even after quenching from 1250°. In the present paper the structure investigation combined

with the texture and magnetic measurements of the CoNiAl alloys with 28 at.% Al have been presented.

Experimental procedure

The $Co_{35}Ni_{37}Al_{28}$ and $Co_{37}Ni_{35}Al_{28}$ (at.%) alloys were cast by induction melting in vacuum using pure elements (more than 99.9 %). After casting, the samples were homogenized in vacuum at 1200°C for 32 hours. The hot rolling was performed at 900°C and resulted in a reduction of sample thickness by 45 % and 90 %. After deformation, the additional heat treatment consisting of annealing at 1250°C for 1 hour and water quenching (WQ) was applied to induce the martensitic transformation. The structure examination was performed in an optical microscope Leica DM IRM microscope equipped with qualitative image analysis software. Martensitic transformation temperatures were determined by differential scanning calorimetry (DSC) using Q1000 TA Instruments calorimeter with cooling and heating rate of 10°C/min. Analytical transmission electron microscopy (TEM) investigations were performed using Tecnai G2 microscope equipped with energy dispersive X-ray analysis system and High Angle Annular Dark Field detector (HAADF). Texture measurements of the deformed sample were performed with the EBSD technique in a scanning electron microscope FEI E-SEM XL-30. The magnetic properties were determined using a vibrating sample magnetometer VSM.

Results and discussion

The set of optical micrographs in Fig. 1 shows the evolution of structure of the $Co_{35}Ni_{37}Al_{28}$ alloy after different treatments. Similar changes of structure were also observed in $Co_{37}Ni_{35}Al_{28}$ alloy.

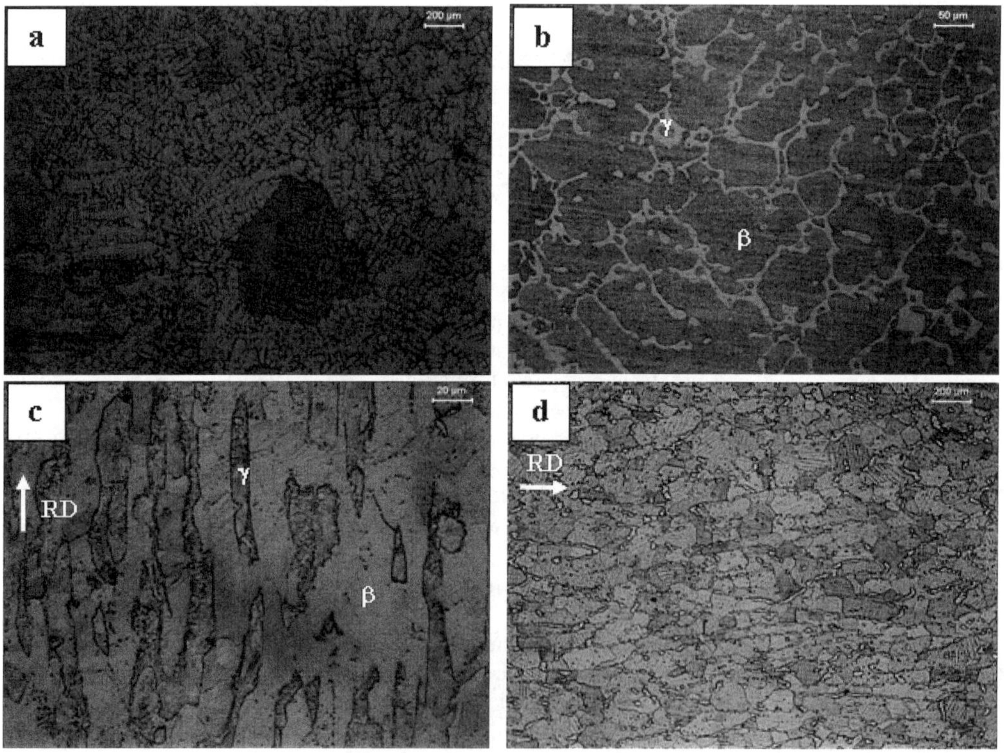

Fig. 1. Optical micrographs of $Co_{35}Ni_{37}Al_{28}$ alloy a) as cast, b) homogenized in vacuum at 1200°C for 32 hours, c) hot rolled 90% at 900°C, and d) hot rolled 90% at 900°C then annealed at 1250°C for 1 hour and WQ.

Characteristic dendritic structure with a mean grain size of 800 μm was observed in the as cast condition, in polarized light. The homogenization process of the ingot at 1200°C for 32 hours was performed in a vacuum furnace. The homogenous samples were used for quantitative analysis of phase volume fraction. Fig. 1b presents a two phase β+γ structure obtained after homogenization. The bright γ phase is mainly distributed at grain boundaries of the dark β phase. Due to the relatively high contrast of the existing phases, a quantitative image analysis was performed, and it helped to establish the volume fraction of γ phase to be 21,5 and 20 % for the $Co_{35}Ni_{37}Al_{28}$ and $Co_{37}Ni_{35}Al_{28}$ alloys, respectively. The values were almost twice lower than those observed earlier in alloys with 25 at.% of aluminium [8]. The hot rolling process was carried out at 900°C in several passes causing a reduction of initial thickness of samples up to 45 % and 90 %. Typical microstructure after the hot rolling is presented in Fig. 1c. The grains of both, γ and β phases elongated parallel to the rolling direction are visible. Additionally, traces of martensite plates can be distinguished in the β grains showing that the martensitic transformation partially occurred during the deformation process. After the hot rolling the samples were annealed at 1250°C for 1 hour and WQ in order to provoke the martensitic transformation. It can be seen in Fig. 1d that the sample is almost fully transformed into the martensite, leaving small amounts of the γ phase unaltered. The martensite grains were of size between 100 to 200 μm and part of them had almost an equiaxial shape, most probably due to the recrystallization process occurred during annealing. However, taking into account the distribution of the γ phase at the boundaries of the β phase it seemed that some texture still remained.

Figure 2a and b presents the DSC curves of the $Co_{35}Ni_{37}Al_{28}$ and $Co_{37}Ni_{35}Al_{28}$ alloys, respectively. The samples were annealed at 1250 °C for 1 hour and then WQ. The temperatures of start martensitic transformation (M_s) are at 98 °C and 50 °C and the reverse one (A_s) at 73 °C and 11 °C for the $Co_{35}Ni_{37}Al_{28}$ and $Co_{37}Ni_{35}Al_{28}$ alloys, respectively. The heat effect produced during martensitic transformation was higher for the $Co_{35}Ni_{37}Al_{28}$ alloy most probably due to a lower volume fraction of the γ phase remaining after the heat treatment. This suggestion will be confirmed in further discussion.

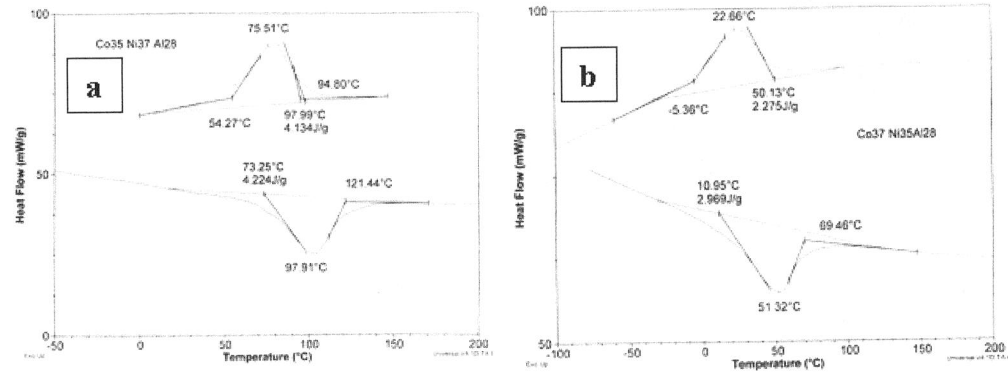

Fig. 2 DSC curves during cooling and heating of a) $Co_{35}Ni_{37}Al_{28}$ and b) $Co_{37}Ni_{35}Al_{28}$ alloys hot rolled and then annealed at 1250°C for 1 hour and WQ.

TEM observations allowed identifying the structure of phases in the investigated alloys, especially at phase boundaries, tetragonality of martensite and its chemical composition. Fig. 3 presents TEM Bright Field (BF) micrographs and corresponding Selected Area Diffraction Patterns (SADP) of the $Co_{37}Ni_{35}Al_{28}$ alloy. The interface boundary between the ordered γ (γ') and $L1_0$ martensite was distinguished by solving the diffraction patterns, which corresponded to the [001]γ' and [112]$L1_0$ zone axes, respectively. The existing martensite was determined to have a tetragonal structure with the lattice parameter: a=b=3.875Å and c=3.139Å (c/a=0.81). The morphology of martensite was identified as lamellar twins often contained within the micro-twins inside the lamellar martensite plates.

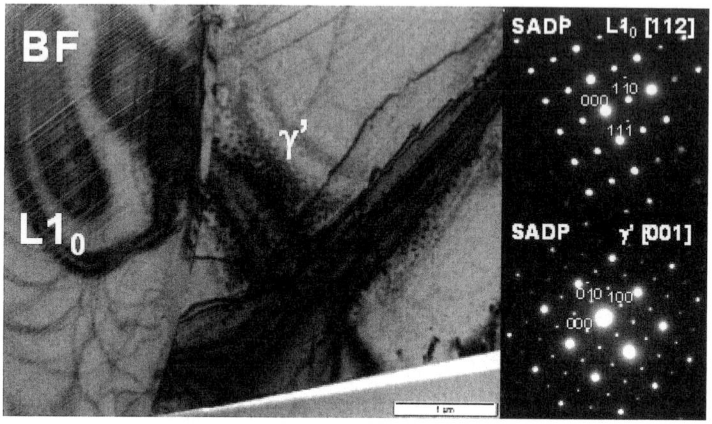

Fig. 3. BF image and corresponding SADP-s of $Co_{37}Ni_{35}Al_{28}$ alloy hot rolled and then annealed at 1250°C for 1 hour and WQ.

Fig. 4 presents a BF image and corresponding SADP of a martensite grain with the micro-twins. The SADP indicates that it is a typical twin orientation $[110]L1_0$ with twins appearing on the $\{111\}L1_0$ planes. The higher magnification of the twinned grain presented as inset of the BF image shows a very thin twin lamellas of width below 20 nm. In order to determine the composition of martensite grains the HAADF combined with EDS techniques were applied.

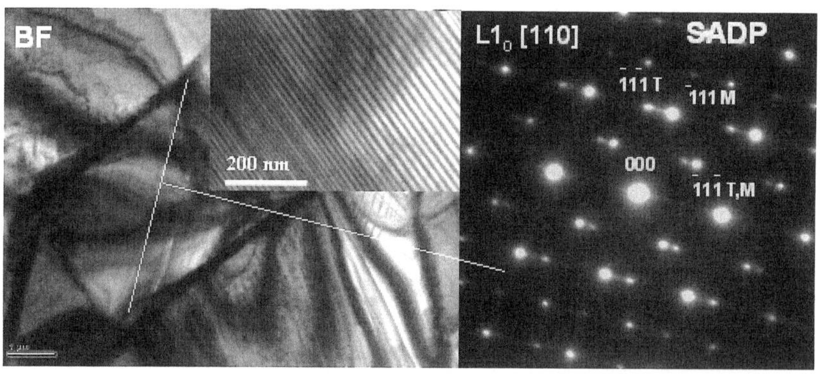

Fig. 4. BF image and corresponding SADP-s of $Co_{35}Ni_{37}Al_{28}$ alloy hot rolled, annealed at 1250°C for 1 hour and WQ.

Fig. 5 shows the BF image and corresponding HAADF image of $Co_{37}Ni_{35}Al_{28}$ alloy after heat treatment. One can see different grains of martensite, with or without micro-twins caused by the orientation effect. The results of chemical point analyses performed in the marked areas, shown in Fig. 5, are presented in Table 1 together with a calculated e/a ratio. One can see a very slight difference in composition of different martensite grains. These differences could be the result of both the channeling effect during the EDS analysis of highly ordered structures or different thickness of individual grains. Additionally, taking into account the standard deviation of the performed analyses being close to ±0.5 at.% it might be concluded that there is no difference in chemical composition of particular martensite grains. The mean e/a ratio calculated using the obtained data from the EDS analyses was determined as e/a=7.725 for both investigated alloys.

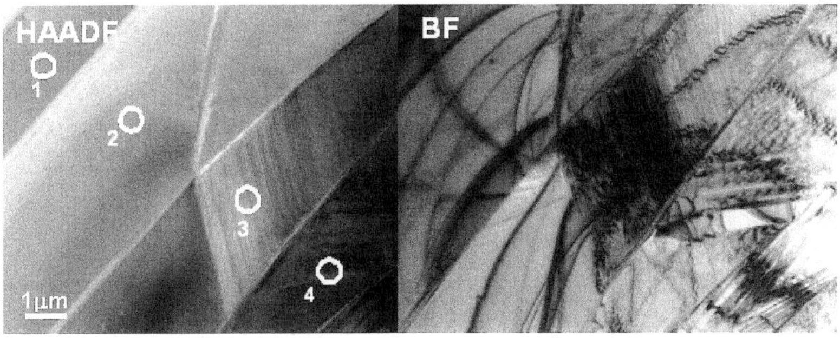

Fig. 5. HAADF image and corresponding BF image of $Co_{35}Ni_{37}Al_{28}$ alloy hot rolled, annealed at 1250 °C for 1 hour and WQ.

Table 1. Results of chemical point analyses performed in marked areas in Fig. 5

At.%	Point1	Point2	Point3	Point4
Co	36,6	36,6	35,8	35,8
Ni	37,1	37,1	37,1	35,9
Al	26,3	26,4	27,2	28,4
e/a	*7.8*	*7.8*	*7.7*	*7.6*

The texture measurements were performed with the EBSD technique of the samples deformed up to 45% of the initial thickness. Fig. 6 a, b presents a SEM micrograph and corresponding phase image obtained as a result of measurements of individual orientations in the $Co_{35}Ni_{37}Al_{28}$ alloy. There were about 8% of unsolved orientations (mostly located in $L1_0$ grains) in both samples, while the volume fraction of the γ phase was 8 and 13% for the $Co_{35}Ni_{37}Al_{28}$ and $Co_{37}Ni_{35}Al_{28}$ alloys, respectively. The pole figure projection of {111} planes (rolling plane) is presented in Fig. 6c and shows the existence of texture component close to the {110} <011> type.

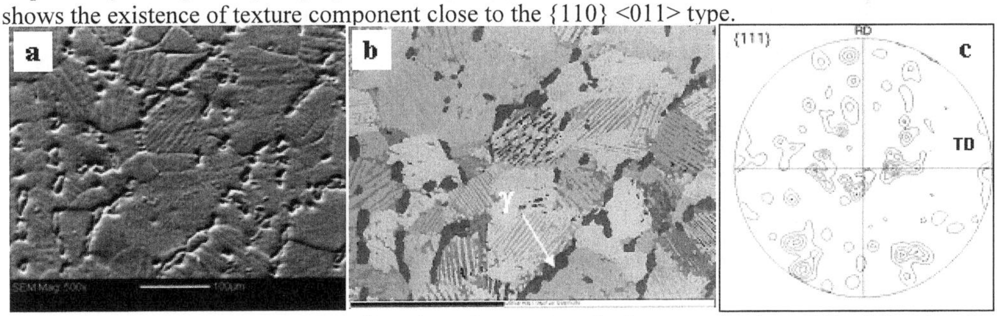

Fig. 6. a) SEM image, b) corresponding phase image, c) pole figure projection of $Co_{35}Ni_{37}Al_{28}$ alloy deformed up to 45% of initial thickness, annealed at 1250°C and WQ.

Unfortunately, in the case of the $Co_{37}Ni_{35}Al_{28}$ alloy the formation of preferred orientations was not observed, most probably due to a higher content of the γ phase which was responsible for plastic deformation during hot rolling. That higher amount of γ phase in the alloy after annealing and WQ was also the reason for a lower heat effect recorded in the DSC curve (Fig. 2) during the martensitic transformation.

The magnetic properties of annealed and WQ alloys were determined by the VSM measurements. Fig. 7 presents the changes of magnetic moment as a function of temperature recorded during heating for both alloys. One can see that magnetic transitions occurred in the temperature range between 50 and 120°C which overlapped with the martensitic transformation temperatures.

Additionally, it can be seen that the magnetic transition (ferromagnetic → paramagnetic) during heating did not occur completely most probably due to the presence of two phase structure $L1_0 + \gamma'$ in the investigated alloys. However, the above observation can be considered as a magneto-caloric effect which will be the subject of further investigations.

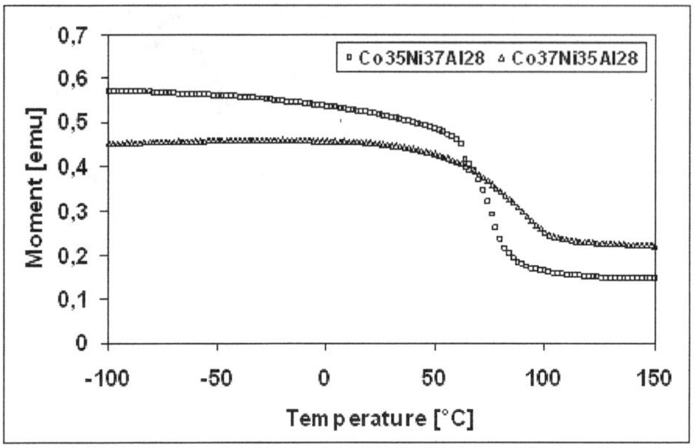

Fig. 7. Magnetic moment as a function of temperature (recorded during heating) for $Co_{35}Ni_{37}Al_{28}$ and $Co_{37}Ni_{35}Al_{28}$ alloys annealed at 1250°C for 1 hour then WQ.

Conclusions

1. The decrease of γ phase content from about 20% after homogenization at 1200°C of ingots down to 8% or 13% after hot rolling at 900°C and additional annealing at 1250°C and WQ was observed in the $Co_{35}Ni_{37}Al_{28}$ and $Co_{37}Ni_{35}Al_{28}$ alloys, respectively.
2. The presence of twinned $L1_0$ martensite with {111}$L1_0$-type twinning plane, tetragonality c/a = 0.81 and the ordered γ' phase was observed using TEM. The chemical point analyses allowed the determination of e/a ratio as 7.725.
3. The texture measurements revealed the existence of texture component close to {110} <011> type but only in the alloy with a lower content (8%) of the γ phase. Its higher content (up to 13%) brought about a decay of texture, most probably due to the localization of plastic deformation inside more ductile γ phase.
4. The magnetization as a function of temperature revealed the overlapping martensitic and magnetic transformation temperatures. The ferromagnetic → paramagnetic transition during heating was not completed due to the existence of two phase structure $L1_0 + \gamma'$ in alloys.

Acknowledgment
This work was supported by The Polish Ministry of Science and Higher Education, grant no. PBZ-KBN-115/T08/04

Literature

1. Stephen A. Wilson at all. Materials Science and Engineering R 56 (2007), p. 1–129
2. K. Ullako, J.K. Huang, V.V. Kokorin, and R.C. O'Handly, Scripta mater. 36, 10, (1997), p. 1333
3. R. Kainuma, M. Ise, C.C. Jia, H. Ohtani, K. Ishida, Intermetallics, 4, (1996), p.151
4 K. Oikawa, L. Wulff, T. Iijima, F. Gejima, T. Ohmori, A. Fujita, et al. Appl Phys Lett 2001;79:3290.
5. V.A. Chernenko, Scripta Materialia 40, (1999), p. 523
6. W. Maziarz, J. Dutkiewicz, L. Tarkowski, R. Santamarta, E. Cesari, Acta Metallurgica Slovaca 13, (2007), p. 160
7. W. Maziarz, Journal of Alloys and Compounds 448, (2008), p. 223
8. W. Maziarz, J. Dutkiewicz, R. Santamarta, E. Cesari, The European Journal of Physics, Special Topics, 158, (2008), p. 137

Solid State Phenomena Vol. 154 (2009) pp 151-156
online at http://www.scientific.net

Effect of annealing conditions on the structure of $Ni_{50}Mn_{29}Ga_{21}$ shape memory alloy

Rafał Wróblewski[1, a], Marcin Leonowicz[1,b]

[1] Warsaw University of Technology, Faculty of Materials Science and Engineering, 141 Woloska Str., 02- 507 Warsaw, Poland

[a]r.wroblewski@inmat.pw.edu.pl, [b]mkl@inmat.pw.edu.pl

Keywords: Ni-Mn-Ga alloys, magnetic shape memory alloys, martensitic transformation.

Abstract. Effect of annealing conditions on the structure and phase transition temperatures for polycrystalline $Ni_{50}Mn_{29}Ga_{21}$ alloy was studied. The final structure strongly depends on the annealing temperature and cooling rate. Ordering of the austenitic phase, performed by annealing at elevated temperatures, plays an important role in formation of the proper martensitic structure. For the $Ni_{50}Mn_{29}Ga_{21}$ alloy annealing below the ordering temperature (1033 K) forms 5M modulated martensitic structure, irrespective of the cooling rate. The 7M modulated martensitic structure can be produced after annealing above the ordering temperature, followed by slow cooling.

Introduction

Ferromagnetic shape memory alloys (FSMA) attract substantial scientific and technological attention due to the giant magnetic-field-induced strain (up to 10%) arising from rearrangement of martensitic variants. This effect was first reported for the Ni_2MnGa Heusler alloys [1]. The NiMnGa alloys undergo upon cooling phase transformation from the parent, cubic $L2_1$ phase to the low symmetry martensite phase. The martensite phase may have various modulated or non-modulated structures, depending on the composition represented by electronic concentration (e/a) [2] and atomic order [3]. The NiMnGa alloys solidify to a B2 disordered phase and upon cooling undergo ordering at around 1053 K by transformation to the $L2_1$ phase [3]. The atomic order of the austenite is supposed to affect the properties of the low temperature martensitic phase. Thermal treatment is necessary to produce the right martensitic structure, together with mechanical treatment (training) which enhances the mobility of twin boundaries [4]. However, the structural features, which change in the course of annealing, are not fully understood. Thus, the effect of heat treatment is currently widely studied. It has been reported for $Ni_{49.5}Mn_{30.5}Ga_{20}$ and $Ni_{52.5}Mn_{24.5}Ga_{23}$ melt-spun ribbon alloys that annealing in the range of 420-670 K promotes increase of the martensitic transformation and Curie temperatures [5]. It has also been found that for $Ni_{50}Mn_{30}Ga_{20}$ alloys annealing at 873 K for 14 h leads to formation of the 7M martensite [6].

In this work we studied the effect of annealing temperature, chosen respectively below and above the ordering temperature, on the structure and transition temperatures for the $Ni_{50}Mn_{29}Ga_{21}$ polycrystalline alloy.

Experimental

The $Ni_{50}Mn_{29}Ga_{21}$ alloy was induction melted and cast invacuum. Afterwards the polycrystalline alloy was homogenized at 1223 K for 48 hours and subsequently furnace cooled slowly. The 25 mm diameter ingot was cut into 3 mm thick slices which were annealed at 873 and 1073 K for 12 hours and subsequently either furnace cooled slowly or quenched in ice water. The composition was checked by EDS technique after each annealing step.

The X-ray diffraction analysis on bulk samples was done using Rigaku MultiFlex diffractometer, Cu Kα radiation.

Magnetic moment was measured on the zero field cooled specimens, on warming using LakeShore 7410 Vibrating Sample Magnetometer (VSM) equipped with a cryostat, at a field of

4 kA/m. Differential thermal analysis (DTA) was done using Labsys DTA/DSC calorimeter with heating rate of 50K/min.

Results and discussion

The transformation temperatures: A_s (martensite-to-austenite start temperature), A_f (martensite-to-austenite finish temperature) and T_c (Curie temperature), recorded on warming, for the specimens annealed at various conditions are shown in Table I and Figure 1. The characteristic temperatures were determined at the deflection points on the M=f(T) curves.

Table I. A_s (martensite-to-austenite start temperature), A_f (martensite-to-austenite finish temperature), T_c (Curie temperatures) [K] and phase constitution, for the $Ni_{50}Mn_{29}Ga_{21}$ alloy after various heat treatments.

Treatment	A_s	A_f	T_c	*Dominant phase*
As-homogenized	335	345	375	5M
873 K slow cooling	335	345	375	5M
873 K quenching	335	340	365	5M
1073 K slow cooling	335	345	370	7M
1073 K quenching	335	340 - 365	365	5M

The A_s points are roughly independent on the annealing temperature and cooling conditions, and are close to 335 K. The A_f temperatures vary slightly for different specimens, however, more substantial differences are shown for the magnetic transformation temperature, T_c, which strongly depends on the atomic arrangement. The T_c is higher for the alloys slowly cooled after annealing, which may indicate better structural ordering. It is consistent with the report [3] that well ordered alloys, where the (0, 0, 0) positions of the $L2_1$ structure are occupied by Mn atoms and the (½, ½, ½) positions by Ga atoms show increased T_c values.

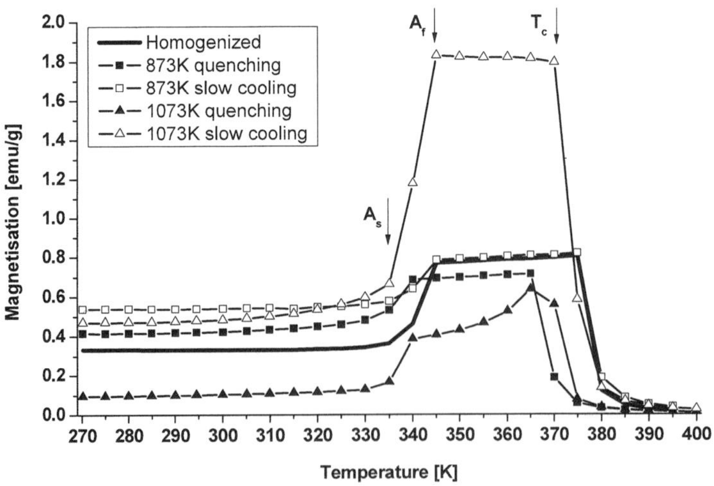

Fig. 1. Magnetisation versus temperature for the $Ni_{50}Mn_{29}Ga_{21}$ alloy after various heat treatments.

Differences in the low field magnetisation values for the martensite and austenite phases, obtained for the specimens having various thermal histories, may also suggest differences in the magnetic anisotropy and structure of these phases, resulting from differences in the ordering, however some influence of local texture in the specimens one can not fully exclude. Especially pronounced are differences in the magnetisation of the ferromagnetic austenite, for the specimens quenched and slowly cooled from 1073 K, respectively (Fig.1). The magnetisation for the quenched specimen grows upon heating in the VSM in the range of 340 K - 365 K giving evidence for atomic rearrangement. On the other hand, after slow cooling from 1073 K, the magnetisation of the ferromagnetic austenite is exceptionally high. A contribution to this high value from good structural ordering and low defect density may explain this behaviour.

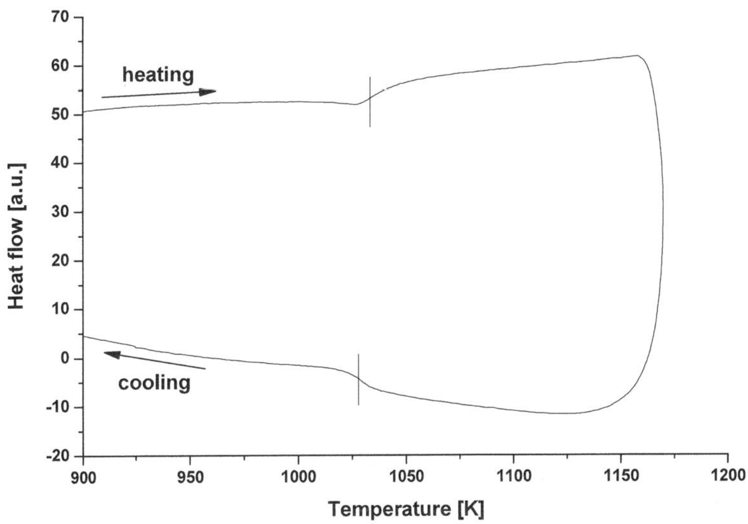

Fig. 2. The DTA trace in the vicinity of the ordering temperature.

The DTA trace (Fig. 2) revealed that the transition from ordered the $L2_1$ to disordered B2 phase occurs on heating at a temperature 1033 K, which is consistent with the published value for similar composition [3,7]. Thus at 1073 K the alloy is in the B2 structure and at 1033 K transforms to the $L2_1$ structure, which on slow cooling provides complete ordering of the austenite phase.

The phase analysis of the specimens was done using X-ray diffraction (XRD) (Fig. 3 Table I). The as-homogenized specimen has a dominant structure of 5M orthorhombic martensite. Annealing at 873 K, irrespective of the cooling rate, does not make any noticeable change to the microstructure and phase constitution.

Both the specimens, slowly cooled and quenched from 873 K, exhibit broadly the same structure, which are characteristic of orthorhombic 5M modulated martensite [9, 7]. This finding was also confirmed by the transmission electron microscopy observations. The TEM microstructures showed martensitic structure containing twin variants (Fig. 4). The selected area electron diffraction pattern (SAEDP) shows that the martensite in this sample is of the 5M type, which is evidenced on the SAEDP by the existence of satellites spaced by 1/5 of the distance between the fundamental maxima.

However, substantial differences are shown for the specimens slowly cooled and quenched from 1073 K, respectively. For the former 7M modulated martensitic structure was obtained [7], whereas for the latter the dominating phase is 5M martensite, however some unidentified peaks are also observed, suggesting a multiphase structure. The calculated pattern was prepared on a basis of orthorombic cell. These studies show that the order-disorder

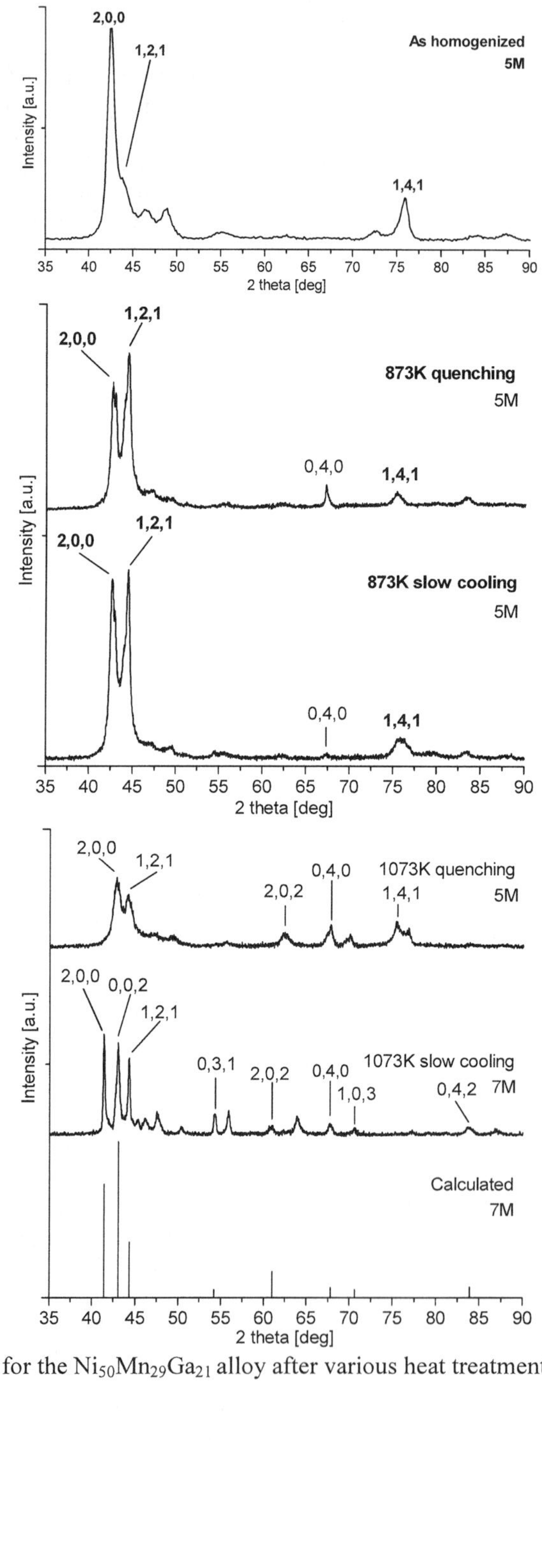

Fig. 3. XRD patterns for the $Ni_{50}Mn_{29}Ga_{21}$ alloy after various heat treatments.

transformation in the austenite plays an important role in formation of the martensitic structure. Quenching from 1073 K apparently restrains complete ordering in the $L2_1$ structure. When slow cooling from 1073 K is applied the diffusion in the austenite enables good ordering of the Ga and Mn atoms and further formation of the 7M martensite.

Fig.4. TEM imaged for the sample containing 5M martensite.

Summary

The results obtained in this study enable to draw the following conclusions:

The final phase structure of the $Ni_{50}Mn_{29}Ga_{21}$ alloy strongly depends on the heat treatment conditions.

Ordering of the austenitic phase, performed by annealing at elevated temperatures, plays an important role in formation of the martensitic structure. For the $Ni_{50}Mn_{29}Ga_{21}$ alloy annealing below the ordering temperature (873 K) forms 5M modulated martensitic structure, irrespective of the cooling rate. The 7M modulated martensitic structure can be produced after annealing above the ordering temperature followed by slow cooling.

Acknowledgement

This study was financed by the Polish Ministry of Science and Higher Education as a Targeted Research Project in the years 2005-2008.

References

[1] Ullakko K., Huang J.K., Kantner C., O'Handley R.C., Kokorin V.V.: Appl. Phys. Lett. 69 (1966) 1966.

[2] Chernenko V.A. : Scripta Mater. 40 (1999) 523.

[3] Sanchez-Alarcos, Recarte V., Pérez-Landazábal J.I., Cuello G.J.: Acta Mater. 55 (2007) 3883.

[4] Gaitzsch U., Pötschke M., Roth S., Rellinghaus B., Schultz L.: Scripta Mater. 57 (2007) 493–495.
[5] Recarte V., Pérez-Landazábal J.I., Gomez-Polo C., Segui C., Cesari E., Ochin P.: Mater. Sci. Eng. A 438-440 (2006) 927.
[6] Gaitzsch U., Pötschke M., Roth S., Mattern N., Rellinghaus B., Schultz L.: J. Alloys. Comp. 443 (2007) 99.
[7] Pons J., Chernenko V.A., Santamarta R., Cesari E.: Acta Mater. 48 (2000) 3027.
[8] Overholser R.W., Wuttig M., Neumann D.A.: Scripta. Mater. 40 (1999) 1095.
[9] Schryvers D. : J. De Physique IV, 5 (1995) C2-225.

Solid State Phenomena Vol. 154 (2009) pp 157-161
online at http://www.scientific.net

Sensor of Current or Magnetic Field Based on Magnetoresistance Effect in $(La_{0.7}Ca_{0.3})_{0.8}Mn_{1.2}O_3$ Manganite Film

V. Dyakonov[1,2, a], S. Piechota[2], K. Piotrowski[1], A. Szewczyk[1], H. Szymczak[1], V. Mikhaylov[2], A. Shemiakov[2], A. Pashchenko[2] and V. Pashchenko[2]

[1]Institute of Physics, Polish Academy of Sciences, Lotnikow 32/46, 02-668 Warsaw, Poland

[2]A.A. Galkin Donetsk Physico-Technical Institute NASU, 83114 Donetsk, Ukraine

[a]dyakon@ifpan.edu.pl

Keywords: Manganite; magnetoresistance effect; film; annular magnet; sensor of current or magnetic field.

Abstract. The main objective of the performed investigations was to enhance sensitivity of a current sensor to weak changes of magnetic field. New design of the sensor of current based on magnetoresistance effect – MRE (MRE = $(R_H - R_0)/R_0$, where R_H is the resistance in magnetic field and R_0 is the resistance without magnetic field) was developed. The sensor was produced in the form of an annular magnet with a gap, in which the $(La_{0.7}Sr_{0.3})_{0.8}Mn_{1.2}O_3$ manganite film possessing large negative MRE was inserted. Nominal current in a controllable electric circuit can change from a few tenths parts of ampere to a hundred of amperes. The limit detectable change of current value depends on the size of gap in the annular magnet. The operation time of sensor at current overload and short circuit is less than 0.3 sec. These magnetoresistors are thermally stable over the temperature range from (- 50 ° C) to (+ 50 ° C). Proposed sensors based on MRE can be applied in many electrical arrangements and devices.

Introduction

The basis for work of the majority of the magnetoresistive sensors of current is the dependence of resistance on magnetic field, i.e., the magnetoresistance effect (MRE). The closest to the proposed and studied sensor of current is the sensor based on the Hall effect [1]. The weak point of such sensor is that the magnetosensible element made of one semi-conductive plate does not allow to extend the limits of measured current in controllable electric circuit. In addition, such sensors have large sizes, are expensive, and their production is labor-intensive.

It is known that the magnetic induction of a magnet in magnetic field is expressed as $B = \mu(H)*H$, where B is the magnetic induction, $\mu = \mu_0\mu_H$ is the magnetic permeability, μ_0 is the permeability of the vacuum, μ_H is the initial magnetic permeability, and H is the intensity of magnetic field.

In systems formed of closed magnetic circuit, increase of sensitivity of sensor to magnetic field could be expected due to the large magnetic induction in toroidal systems [2].

The ferrites possessing the large magnetic induction and showing MRE [3, 4] were applied in the first sensors of current as the magnetosensible elements [2]. Magnetoresistors were produced in the shape of a ring with silver contacts plotted on the opposite sides of the ring. The ring was made of manganese-zinc ferrites with different magnetic permeability. Previous studies showed that the value of magnetosensitivity $d(\Delta R/R_0)/dH$, for example in the sample of $Mn_{0.9}Zn_{0.1}Fe_2O_4$, exceeds magnetosensitivity of the sensors based on bulk ferrites up to 100 times. However, a small magnetosensitivity of the toroidal ferrite magnetoresistors to magnetic field of the order of a few hundred Oe was their weak point.

The main objective of the performed investigations was to develop a new design of the sensor of current based on magnetoresistance effect and to enhance the magnetosensitivity of the sensor to weak changes of magnetic field.

Results

New design of the sensor of current, being in fact the sensor of magnetic field produced by this current, based on MRE was developed and studied. The sensor was produced in the form of an annular magnet with a gap, made of permalloy, i.e. of the material characterized by high magnetic induction in small magnetic field. To improve sensitivity of the sensor, the magnetosensible element from the manganite film was inserted into the gap of the magnet. At present, the largest negative MRE is observed in manganese - lanthanum perovskites of the composition $La_{0.7}Me_{0.3}MnO_3$ (Me = Ca, Sr).

The application of manganite films is based on a strong dependence of their resistance and magnetoresistance on magnetic field. The electric resistance, ρ(T), becomes smaller (Fig. 1) and magnetoresistance effect increases (Fig. 2) with increasing magnetic field.

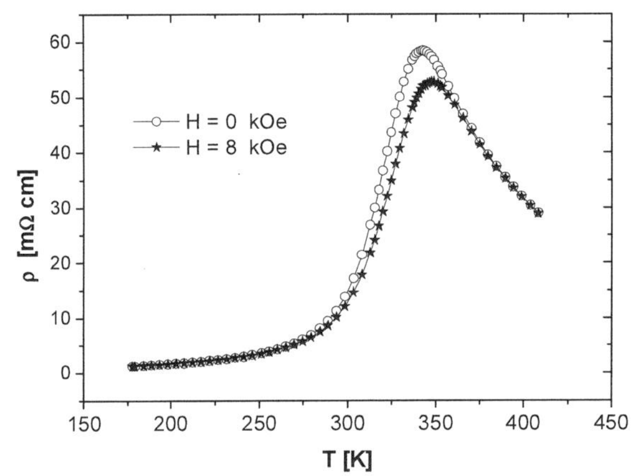

Fig. 1. Temperature dependence of resistance of $(La_{0.7}Sr_{0.3})_{0.8}Mn_{1.2}O_3$ film at H = 0 and 8 kOe

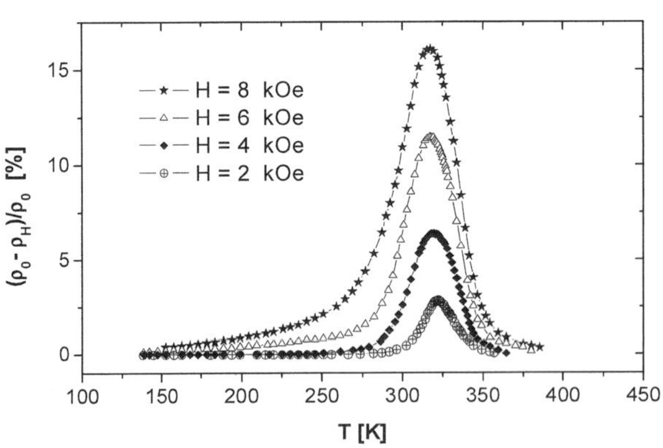

Fig. 2. Temperature dependence of magnetoresistance of $(La_{0.7}Sr_{0.3})_{0.8}Mn_{1.2}O_3$ film

The application of manganese - lanthanum films with high MRE is the main distinction between the proposed sensor of current and the sensors based on ferrite rings and Hall effect.

The proposed sensor of electric current is illustrated schematically in Fig. 3.

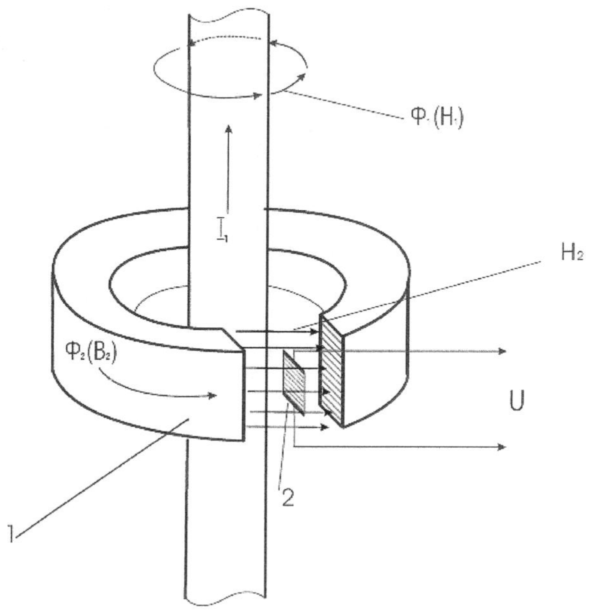

Fig. 3. Diagram of the sensor of electric current

It consists of the annular magnet (1) from permalloy, which is located around the current conductor with the measured current, and magnetosensitive element (2) produced from the manganese-lanthanum-strontium $(La_{0.7}Sr_{0.3})_{0.8}Mn_{1.2}O_3$ film of the thickness of 1000 Å and placed in the gap of the magnet. The manganite with the excess of manganese is used, because it possesses both the high metal-insulator transition temperature (above room temperature) and the large negative MRE.

The sensor works as follows: when the current flows through the current conductor, the magnetic field of the intensity H_1 is induced around it, which produces the magnetic flux of Φ_1 penetrating the annular magnet (1). As a result, the magnetic field of the intensity H_2 occurs in the gap of the magnet and it is recorded by the magnetosensitive element due to MRE. This signal is compensated by the voltage produced in compensating coil with an electronics. The current of about 1 A flowing through the current conductor produces the magnetic field of about 0.5 T in the 1 mm gap of the annular magnet, and the magnetoresistance effect induced by it achieves about twenty percentages. The limits of current flowing through the conductor, at which the sensor works, can change from several mA to several hundred ampere. The sensor sensitivity to current variation (dMR/dH%/(A/m)) for the low and upper current limits is established to change from 1-2%(A/m) to 15-20%/(A/m), respectively. The operation time of sensor, i.e., the time needed by electronics to disconnect the system from the electricity supply at current overload and short circuit, was established to be less than 0.3 sec. These magnetoresistors are thermally stable over the range of temperatures from (-50°C) to (+50°C).

In Photo 1, the design of the sensor of electric current involving the annular magnet, the magnetoresistive $(La_{0.7}Sr_{0.3})_{0.8}Mn_{1.2}O_3$ film and the compensating coil is present.

Photo 1. Design of the sensor of electric current

In Photo 2, the sensor of electric current in conjunction with the Wheatstone bridge converting resistance into voltage is shown.

Photo 2. Sensor with Wheatstone bridge.

Summary

High magnetosensitivity, large limit of measured currents, lack of moving electric contacts, high reliability and practically unrestricted time of work allow to use the proposed sensor of current with manganese-lanthanum-strontium films as a magnetosensitive element in many electrical arrangements and devices.

This sensor of current based on MRE can find application in non-contact electric devices of constant currents, electric transducers, relay systems, electrical cars and train, as well as in devices of automatic disconnection of the power supply system to protect people from the electric current effects (e.g., in mines, chemical divisions, oil stations, devices for protection of refrigerators and washing machines and others).

This work was financed in part by the Polish Ministry of Science and Education as a Targeted Research Project, over the period 2005–2008 (Project PBZ-KBN-115/T08/01).

References

[1] G.Ja. Portnoj and O.A. Postnuch : Devices and Systems of Control No. 6 (1992), p. 11.

[2] V.P. Pashchenko, M.I. Nosanov and O.A. Shemiakov : Declarative Patent for Use, Ukraine, UA 45153 Bulletin No. 3 (2002).

[3] K.P. Belov, A.N. Goriaga, V.N. Pronin and L.A. Skipertova : JETF Lett. 37 (1983), p. 392.

[4] E. Rezlesku and N. Rezlesku : J. Magn. Magn. Mater. 193 (1999), p. 501

Solid State Phenomena Vol. 154 (2009) pp 163-168
online at http://www.scientific.net

Field Dependence of the Refrigerant Capacity for $La_{0.6}Ca_{0.4}MnO_3$ Manganite

R.Szymczak[1,a], A.Kolano-Burian[2,b], R.Kolano[2,b], R.Puzniak[1,a], V. P. Dyakonov[1,a,3], E. E. Zubov[3,c], O. Iesenchuk[1,a] and H. Szymczak[1,a]

[1]Institute of Physics, Polish Academy of Sciences, Lotnikow 32/46, 02-668 Warsaw, Poland

[2]Institute of Non-Ferrous Metals, Sowinskiego 4, 44-101 Gliwice, Poland

[3]Donetsk Institute for Physics and Engineering, National Academy of Sciences of Ukraine, R. Luxemburg Street 72, 83114 Donetsk, Ukraine

[a] szymh@ifpan.edu.pl [b] olak@imn.gliwice.pl [c] zubov@dyakon.fti.ac.donetsk.ua

Keywords: Manganite; magnetocaloric effect; magnetic refrigeration; magnetic cooling; refrigerant capacity; magnetic entropy.

Abstract. The magnetocaloric effect in $La_{0.6}Ca_{0.4}MnO_3$ manganite has been investigated. The isothermal magnetization versus applied magnetic field at various temperatures in the vicinity of Curie temperature was measured, and the temperature dependence of magnetic entropy change was determined using Maxwell's relation. This value is comparable to that in Gd. Nevertheless, the relative cooling power of $La_{0.6}Ca_{0.4}MnO_3$ was shown to be considerably lower than that of Gd. The experimental results have been analyzed in frames of a phenomenological statistical model. This model considers explicitly Jahn-Teller interactions and allows prediction of the field dependences of the magnetic entropy change and the relative cooling power.

Introduction

In recent years, there has been increasing interest in using manganities not only as material having colossal magnetoresistivity but also as material having promising magnetocaloric properties (see review paper [1]). The magnetocaloric effect arises due to existence in magnetic materials two energy reservoirs with phonon and magnon excitations. These two reservoirs are coupled by the spin-lattice (in other words – magnetoelastic) interactions. An external magnetic field affects the spin degrees of freedom resulting in heating or cooling of the magnetic material. The magnetocaloric effect increases with increase of the applied magnetic field and with the change of magnetization. It means that this effect reaches its maximum in the vicinity of magnetic phase transitions.

A large magnetocaloric effect makes manganities excellent candidates for working materials in magnetic refrigeration units especially because these are less costly than any other materials, particularly commonly used materials based on gadolinium. Moreover, we have shown [2] that the magnetic entropy change $|\Delta S_M|$ in La-based manganites is larger than that exhibited by gadolinium. It has to be stressed that the possibility of tuning the working temperature with minor compositional change, their extremely small magnetic hysteresis, and their enhanced electrical resistivity (similar to that observed in ferrites) which would reduce eddy current losses are all beneficial characteristics for their application as magnetic refrigerators.

The criterion of maximum $|\Delta S_M|$ is, of course, not the only one. According to [3] refrigerant capacity defined as $RC = |\Delta S_M| \cdot \Delta T_{cyc}$ ($\Delta T_{cyc} = T_{hot} - T_{col}$, where T_{hot} and T_{col} are the upper and the lower operating temperatures in the cycle) should reach its maximum value. Such defined parameter RC is a compromise between the magnitude of the magnetic entropy change and the width of the peak. Moreover, it is a suitable metrics for comparing the performance of different materials. However, the most important parameter for the magnetic refrigerators with a limited high magnetic field volume is the relative cooling power (RCP) [4,5] defined as

$$RCP(S) = |\Delta S_M| \delta_{FWHM}, \qquad (1)$$

where δ_{FWHM} is the full width at half maximum of the magnetic entropy change $|\Delta S_M(T)|$ curve. This parameter will be used in this paper when discussing refrigerant capacity of the magnetic materials.

All of the parameters related to magnetocaloric effect such as the magnetic entropy change, refrigerant capacity and relative cooling power depend strongly on the magnitude of the applied magnetic field H. The knowledge of their field dependences is of great importance since it provides tools for making plausible extrapolations and interpolations to magnetic-field values necessary for construction of magnetocaloric refrigerators. Generally, the mentioned parameters increase with increasing maximum applied field H. The first explanation of the field dependence of $|\Delta S_M|$ was given by Oesterreicher and Parker [6]. They have shown that in mean field approximation at the Curie temperature $|\Delta S_M| \propto H^n$, where n=2/3 (see also [7]). Franco et al. [8] have studied the field dependence of $|\Delta S_M|$ in materials with a second order phase transition in three characteristic regions: well below the Curie temperature (T_C), at T_C, and well above T_C. According to their calculations for temperatures well below T_C, n = 1 and well above T_C, n = 2. To determine $n(T_C)$ the Arrott-Noakes equation of state [9] was used. It leads to the following expression:

$$n(T_C) = 1 + \delta^{-1}(1 - \beta^{-1}), \qquad (2)$$

where δ and β are the critical exponents. In the case of a mean field model $\delta = 3$, $\beta = ½$ and consequently $n(T_C) = 2/3$.

In the same approximation the relative cooling power (RCP) depends on applied field H according to the law H^n with $n = 1 + \delta^{-1}$ [10].

In the present paper, we report experimental results on magnetic field dependence of the magnetic entropy changes in $La_{0.6}Ca_{0.4}MnO_3$ manganite with a second order phase transition, and the results are discussed in terms of recently developed theoretical models.

Experimental results

A ceramic sample $La_{0.6}Ca_{0.4}MnO_3$ was synthesized by the method described previously [2]. Analysis of the X-ray powder diffraction patterns confirmed the formation of the perovskite-type structure and showed that the samples were single phase. A Quantum Design Physical Property Measurement System (PPMS) was used for magnetic measurements. The magnetization was measured with a field steps of 3 kOe in a range of 0 - 90 kOe and with temperature steps of 1-2 K over a temperature range around T_C. The changes of field and temperature were sufficiently slow to consider all magnetization processes as isothermal. To calculate the magnetic entropy change the measurements of M(H) isotherms in a wide temperature interval (from 310 to 210 K) have been employed. For this Maxwell relations were used [2].

The temperature dependence of magnetization measured in different magnetic fields is shown in Fig.1. The Curie temperature, defined by the maximum in the value of $|dM/dT|$ is about 260 K. It is seen that paramagnetic-ferromagnetic transition is very narrow but no hysteresis is observed and this transition is identified as second-order one. Isothermal magnetization measured in the vicinity of T_C

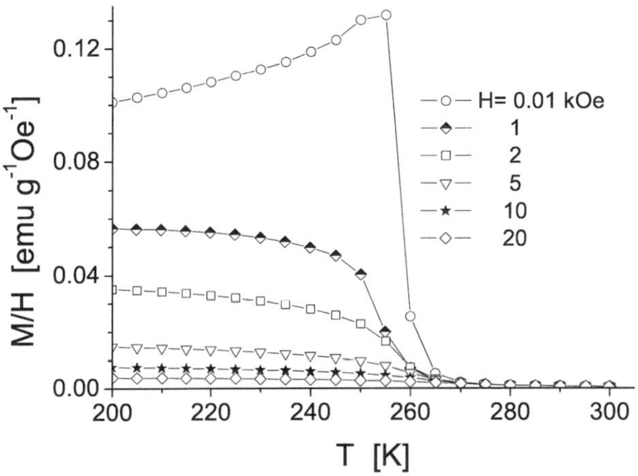

Fig.1. Temperature dependence of magnetization measured in different magnetic fields for $La_{0.6}Ca_{0.4}MnO_3$

were used to determine ΔS_M following standard procedure described in [2]. Figure 2 shows the magnetic entropy change as a function of temperature in various magnetic fields. It is seen that

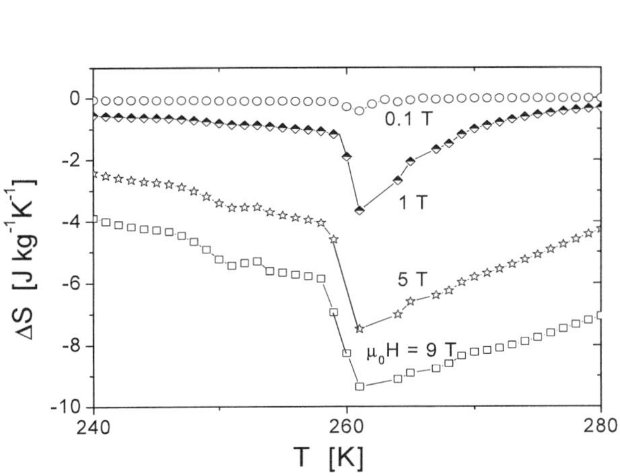

Fig.2. Magnetic entropy change as a function of temperature in various magnetic fields for $La_{0.6}Ca_{0.4}MnO_3$

magnetocaloric effect in $La_{0.6}Ca_{0.4}MnO_3$ is very high and reaches value observed in Gd (for Gd ΔS_M=-10.2 J/kgK for μ_0H = 5T [1]). Using the results presented in Fig.2 the relative cooling power may be determined. The results are shown in Fig.3 as a function of magnetic field H. It may be shown that the RCP scales with magnetic field as RCP $\propto H^{1.35}$. Although maximum magnetic entropy change in $La_{0.6}Ca_{0.4}MnO_3$ reaches value observed in Gd, nevertheless the RCP of this sample is, unfortunately, considerably smaller than that of Gd (for Gd RCP=410 for for $\mu_0H = 5T$ [1]). It has to be mentioned that according to [10] the RCP for $Fe_{78}Co_5Zr_6B_5Ge_5Cu_1$ amorphous alloys scales with field as $H^{1.23}$.

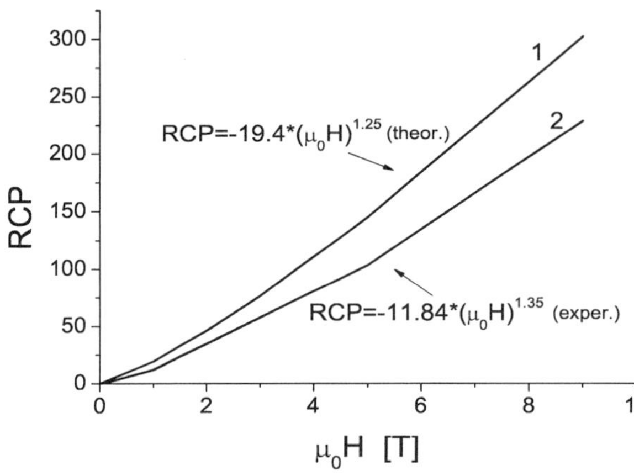

Fig.3. Scaling law describing the field dependence of the refrigerant capacity (theory- curve 1; experiment –curve 2) for $La_{0.6}Ca_{0.4}MnO_3$

Discussion

The experimental results will be analyzed in frames of a phenomenological statistical model developed by Biernacki [11-13]. It is a two-level model describing specific behavior of the e_g electrons in manganites. The model contains two adjustable parameters derived from the experimental value of T_C.

.It has to be stressed that developed model describes properly fundamental properties of manganites, including isotope effect. Important element of the model is that it considers explicitly electron-phonon interaction. According to the model the free energy of the system can be expressed as:

$$F = \varepsilon n^2 - \mu H - k_B T \ln \{1 + 2 \exp [- (tR - 2\varepsilon n + 2\mu H)/k_B T)]\}, \quad (3)$$

where $n \in [0,1]$ is the e_g electron distribution function, tR is the product of the hopping integral t between Mn ions multiplied by the Huang-Rhys factor R, ε is the Jahn-Teller energy, μ is the magnetic moment.

The electronic entropy S can be obtained from the eq. (3):

$$S (H,T) = - k_B [-n \ln 2 + n \ln n + (1 - n)\ln(1 - n)]. \quad (4)$$

The above expression was used to calculate the magnetic entropy change as a function of temperature for $La_{0.6}Ca_{0.4}MnO_3$. In the calculations one fitting parameter namely $T_C = 260$ K was applied. The calculated values of magnetic entropy change $|\Delta S_M|$ (Fig.4) are considerably higher than those measured in experiment (Fig.2). It may be due to intrinsic (phase separation) and/or

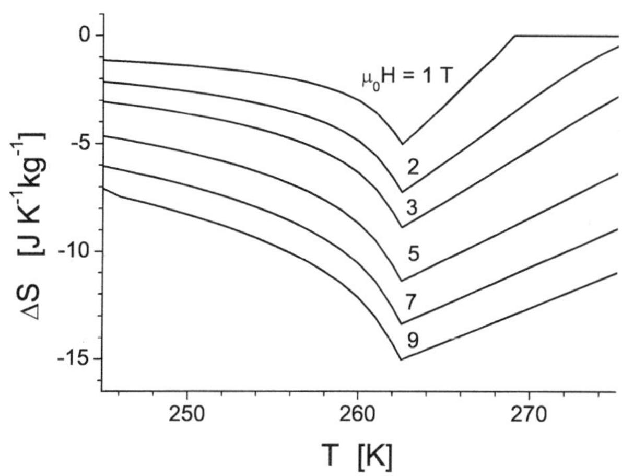

Fig.4. Theoretical calculation of the magnetic entropy change as a function of temperature for $La_{0.6}Ca_{0.4}MnO_3$

extrinsic inhomogeneities typical of manganities. The results presented in Fig.4 were applied in calculations the exponent n characterizing the field dependence of ΔS_M (Fig.5) and RCP (Fig.3). It is seen that the model satisfactorily describes experimental observations. The character of n(T) curve is also in agreement with experimental data reported for other magnetocaloric materials (see e.g. [14,15]).This simple two-parameter model indicates that electron-phonon coupling determines the temperature dependence of ΔS_M and RCP in manganites.

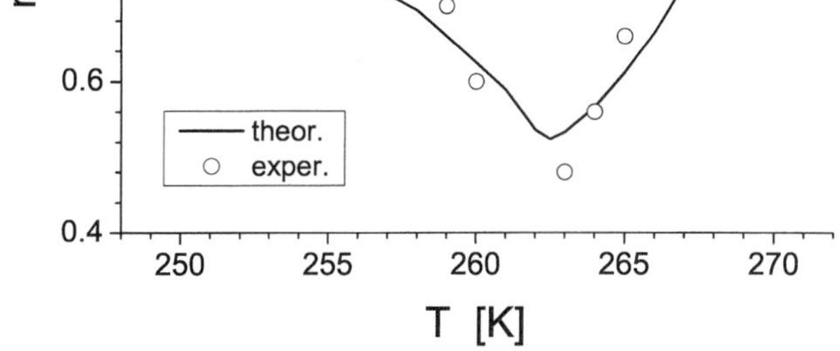

Fig.5. Experimental and theoretical temperature dependences of the exponent characterizing the field dependence of ΔS_M for a $La_{0.6}Ca_{0.4}MnO_3$ manganite.

Summary

In conclusion, the magnetocaloric effect in $La_{0.6}Ca_{0.4}MnO_3$ manganite has been studied. The obtained results show a large magnetic entropy change at the Curie temperature. The magnitude of magnetic entropy change for this manganite is comparable to the value observed in Gd. However, the relative cooling power of $La_{0.6}Ca_{0.4}MnO_3$ was shown to be considerably lower than that of Gd. The experimental results have been analyzed in frames of a phenomenological statistical model developed by Biernacki. The model considers specific behavior of the e_g electrons and contains only one adjustable parameter which is the experimental value of T_C. The calculated values of the magnetic entropy change $|\Delta S_M|$ (Fig.4) and the relative cooling power are considerably higher than those measured in experiment probably because of intrinsic and extrinsic inhomogeneities typical of manganities.

Acknowledgements

This work was partly supported by Ministry of Science and Higher Education of Poland Project PBZ-KBN-115/T08/2004

References

[1] M.-H. Phan and S.-C. Yu: J.Magn.Magn.Mater. Vol. 306 (2007),p.325

[2] R.Szymczak, M.Czepelak, R.Kolano, A. Kolano-Burian, B.Krzymanska and H.Szymczak: J.Mater.Sci. Vol.43 (2008), p.1734

[3] M.E. Wood and W.H.Porter: Cryogenics Vol.25 (1985), p.325

[4] K.A. Gschneidner , Jr. and V.K.Pecharsky: Ann. Rev. Mater. Sci. Vol.30 (2000), p.387

[5] J.Yang and Y.P.Lee: J.Appl.Phys. Vol.102 (2007), p.033913

[6] H.Oesterreicher and F.T. Parker: J.Appl.Phys. Vol.55 (1984), p.4334

[7] Q.Y. Dong, H.W. Zhang, J.L.Shen, J.R. Sun and B.G. Shen: J. Magn. Magn. Mater. Vol.319 (2007), p.56

[8] V.Franco, J.S. Blazquez and A.Conde: Appl. Phys. Lett. Vol.89 (2006), p.222512

[9] A.Arrott and J.E.Noakes: Phys.Rev.Lett. Vol.19 (1967), p.786

[10] V.Franco, A.Conde, J.M. Romero-Enrique and J.S. Blazquez : J.Phys.:Condens.Matter Vol. 20 (2008), p.285207

[11] S.W. Biernacki: Solid State Communication Vol.116 (2000), p.471

[12] S.W. Biernacki: Phys.Rev. B Vol.66 (2002),094405, Vol.68 (2003), p.174417

[13] S.W. Biernacki and H.-J. Schulz: Phys.Rev. B Vol.70 (2004), p.092405

[14] V.Franco, J.S. Blazquez and A.Conde: J.Appl.Phys. Vol 100 (2006), p.064307

[15] V.Franco, J.S. Blazquez, M.Milan, J.M. Borrego, C.F.Conde and A.Conde: J.Appl.Phys. Vol 101 (2007), p.09C503

Solid State Phenomena Vol. 154 (2009) pp 169-173
online at http://www.scientific.net

Magnetic Anisotropy of Nanocrystalline HITPERM-Type Alloys and Its Correlation with Application

Jarosław Ferenc[1,a], Maciej Kowalczyk[1,b], Tatiana Erenc-Sędziak[1,c], Xiubing Liang[2,d], Gabriel Vlasák[3,e] and Tadeusz Kulik[1,f]

[1]Faculty of Materials Science and Engineering, Warsaw University of Technology, ul. Wołoska 141, 02-507 Warsaw, Poland

[2]National Key Laboratory for Remanufacturing, Academy of Armored Forces Engineering, Dujiakan 21, Changxindian, 100072 Beijing, China

[3]Institute of Physics, Slovak Academy of Sciences, Dúbravská cesta 9, 845 11 Bratislava, Slovakia

[a]jferenc@inmat.pw.edu.pl, [b]mkowalczyk@inmat.pw.edu.pl, [c]terenc@inmat.pw.edu.pl, [d]liangxiubing@yahoo.com.cn, [e]fyzivlas@savba.sk, [f]tkulik@inmat.pw.edu.pl

Keywords: nanocrystalline materials, magnetically soft alloys, magnetostriction, magnetocrystalline anisotropy, magnetoelastic anisotropy.

Abstract. Structure as well as magnetic and magnetoelastic properties of nanocrystalline (Fe,Co)-(Hf,Zr)-Cu-B alloys (HITPERM-type) were investigated in order to find out which factors are responsible for the magnetic hardening of these magnetically soft materials. Magnetoelastic anisotropy, caused by the presence of cobalt, was found to play the predominant role in the observed increase of coercive field. On this basis, guidelines on chemical composition and crystallisation process selection were suggested for two fields of application: soft magnetic cores and sensors or actuators cores.

Introduction

Amongst many materials that exhibit magnetic softness, i.e. may be magnetised and remagnetised with weak magnetic field, there are nanocrystalline alloys, such as FINEMETs (Fe-Si-Cu-Nb-B) [1], NANOPERMs (Fe-Zr-Nb-Cu-B) [2] and HITPERMs (Fe-Co-Zr-Nb-Cu-B) [3]. Changes of magnetic state take place with the magnetic domain wall movement and orientation of magnetic moments within the domains. Low magnetic anisotropy is therefore necessary to achieve magnetic softness. There are two main components of magnetic anisotropy: magnetocrystalline and magnetoelastic ones. Magnetocrystalline anisotropy may be greatly reduced via the mechanism of averaging out of randomly oriented moments of single nanocrystals [4]. Magnetoelastic anisotropy should be as low as possible, if we seek materials for soft magnetic applications, such as cores of transformers, chokes, etc. In this case, precise control of structure is necessary to obtain proper balance between the magnetostriction of crystalline and amorphous phases constituting the material. However, for e.g. force sensors and actuators it is more convenient to apply materials that exhibit high magnetostriction, so they must have a larger magnetoelastic anisotropy. In this paper, the correlation between structure, magnetic anisotropy and applications of nanocrystalline alloys is discussed taking HITPERM-type alloys as an example.

Experimental

Alloys, with nominal composition of $(Fe_{0.5}Co_{0.5})_{93-x}(Hf_{1-v}Zr_v)_xCu_1B_6$ ($x = 7$ and 9, $v = 0$ and 0.5) and $(Fe_{1-z}Co_z)_{86}Hf_7Cu_1B_6$ ($z = 0 - 1$) were produced by arc melting of pure elements and Fe-B prealloy. The ribbons of amorphous alloys were made by melt spinning, and, subsequently, were subjected to isothermal heat treatment for 1 hour at 550°C in order to obtain nanocrystalline alloys. The structure of the alloys was investigated by x-ray diffraction (XRD) and transmission electron microscopy

(TEM). Hysteresis loops were recorded with the quasistatic hysteresis loop tracer [5], and magnetostriction coefficient was measured with the capacitance method [6].

Results

For the investigated HITPERM alloys, soft magnetic properties are preserved in the nanocrystalline state – their coercivity does not exceed 30 A/m. This is obvious considering the condition that the coercivity is proportional to 6th power of grain size [4], and the studied alloys exhibit average grain size not larger than 30 nm (see Fig. 1a). However, although there is a large difference between grain size of the alloys z = 0.5 and z = 0.6 or 0.8 (D_{avg} = 15.7, 28.0 and 26.2 nm respectively), there is a little difference between their coercivity (H_c = 25, 26 and 30 A/m respectively) – see Fig. 1c. This result may mean that the condition of soft magnetism of the nanostructured alloys is well met, regardless of the mean grain size being 15.7 nm or almost twice larger. Therefore, magnetocrystalline anisotropy is not key for the studied HITPERM alloys. What is important here, is the magnetoelastic anisotropy. When looking at the dependence of the saturation magnetostriction coefficient (λ_s) on cobalt content (Fig. 1b) one can notice that λ_s reaches maximum for $z = 0.5$. Similar behaviour of coercivity on z is observed for $z < 0.8$. This correlation indicates that for low Co content (z up to 0.5), magnetoelastic anisotropy prevails and is the main factor influencing the magnetic behaviour of the investigated alloys. The value of magnetostriction coefficient depends on cobalt content, so for the magnetically soft Fe based alloys the addition of Co is detrimental. Obviously, cobalt dope can not be avoided, because it ensures the desired soft magnetism at high temperature as compared to NANOPERM alloys.

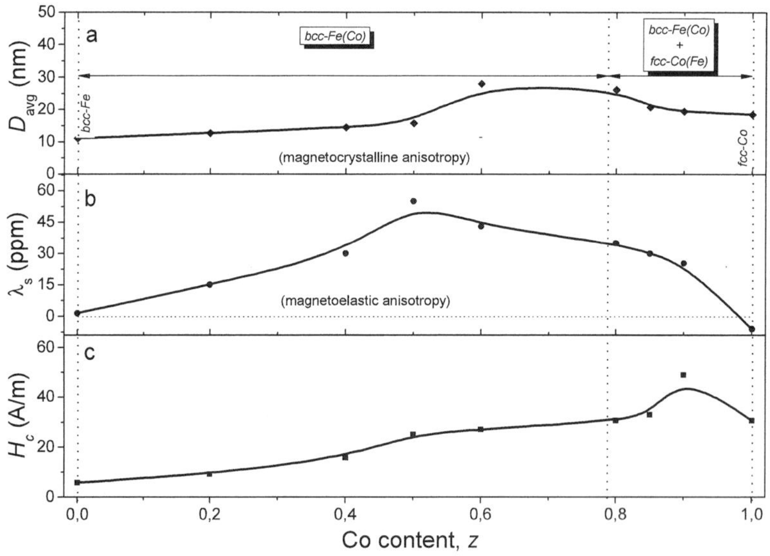

Fig. 1. Dependence of a) average grain size of crystalline phase(s), b) coefficient of magnetostriction and c) coercive field on the cobalt content of nanocrystalline $(Fe_{1-z}Co_z)_{86}Hf_7Cu_1B_6$ (z = 0 - 1) alloys.

The importance of magnetoelastic anisotropy may be further confirmed by the experiment with ageing of nanocrystalline HITPERM alloys by long-lasting annealing at 500 or 550°C [7]. There were two optimally nanocrystallised alloys: $(Fe_{0.5}Co_{0.5})_{86}Hf_7Cu_1B_6$ and $(Fe_{0.5}Co_{0.5})_{84}$-$(Hf_{0.5}Zr_{0.5})_9Cu_1B_6$. In the as-nanocrystallised state, their coercivity was 22 and 41 A/m respectively, but after prolonged annealing (e.g. for 3000 hours at 500°C), H_c was of 44 and 107 A/m respectively. This increase of coercivity might be caused by the increase of magnetic anisotropy and/or by the appearance of secondary phases. In fact, secondary phases were present in the structure of both alloys after annealing, but their amount was smaller than it could harden the materials (as compared to the samples overheated during nanocrystallisation – see Fig. 2). From this we conclude that the reason for the increase was the higher value of rather magnetocrystalline anisotropy than of magnetoelastic anisotropy. The average grain size and crystalline volume fraction of the stable alloy, containing 7% of Hf, and the instable one, containing 9% of (Hf+Zr), before and after prolonged annealing were compared. In the case of the stable alloy, a tiny increase of volume fraction and size of primary crystals was found, while for the instable one we observed a significant increase of volume fraction and size of primary crystals [7]. Considering the observation made in the previous paragraph that the grain size is not important, as long as grains are smaller than certain threshold, one may deduce that volume fraction of crystals may be important. Indeed, this value has a major impact on the effective magnetostriction coefficient, which depends on the proportion of volumetric fractions of crystalline and amorphous phases. In the optimally nanocrystallised state, this proportion assures low value of λ_S. After ageing, if a material is susceptible to the continuation of crystallisation, this proportion goes out of the optimum and the effective λ_s is significantly higher. This clearly leads to the larger value of magnetoelastic anisotropy and, finally, to the worsening of magnetic softness.

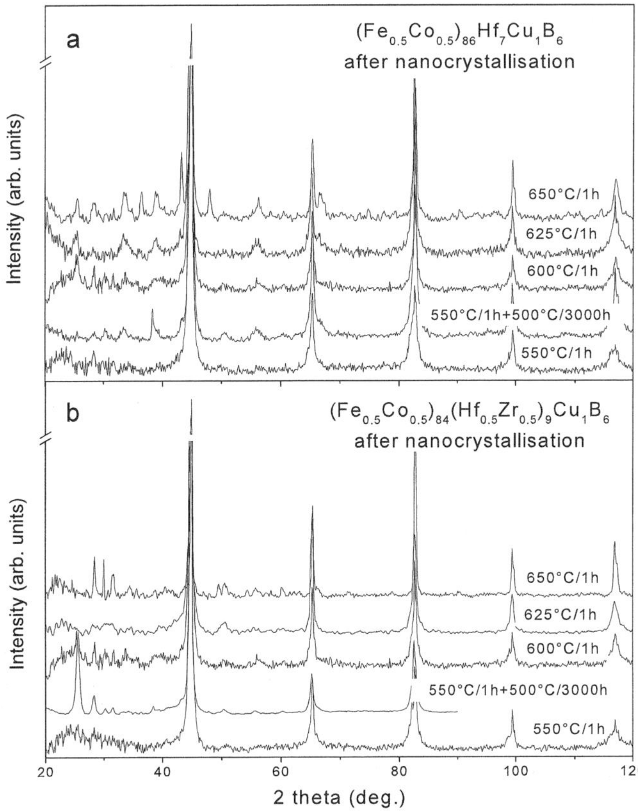

Fig. 2. XRD patterns of nanocrystalline a) $(Fe_{0.5}Co_{0.5})_{86}Hf_7Cu_1B_6$ and b) $(Fe_{0.5}Co_{0.5})_{84}(Hf_{0.5}Zr_{0.5})_9Cu_1B_6$ alloys after nanocrystallisation at various temperatures and after long-term annealing (500°C for 3000 hours).

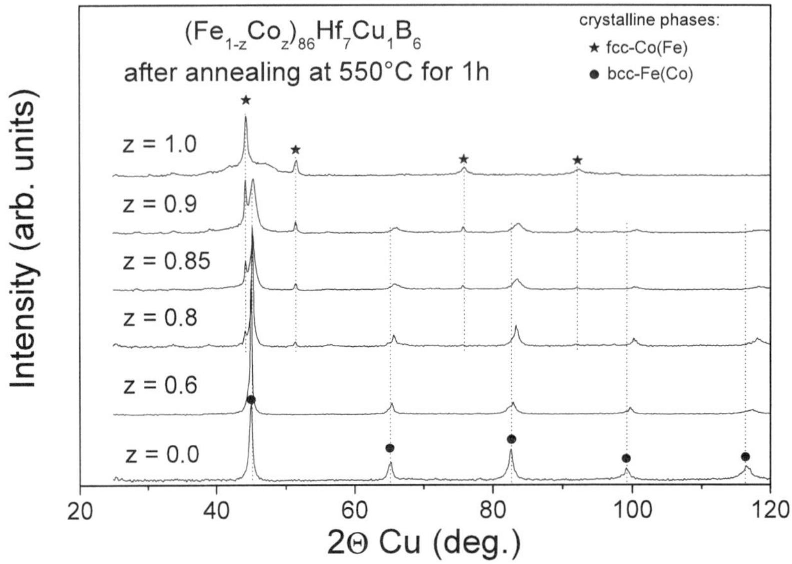

Fig. 3. XRD patterns of nanocrystalline $(Fe_{1-z}Co_z)_{86}Hf_7Cu_1B_6$ ($z = 0$ - 1) alloys after nanocrystallisation at various temperatures and after long-term annealing (500°C for 3000 hours).

The above presented argument may suggest that in the HITPERM alloys, the structure, if nanometres-sized grains are present, is of a lesser importance. It is true, but there is a noticeable discrepancy in the combined plot of dependence of mean grain size, D_{avg}, and coercivity, H_c, on cobalt content, z (Fig. 1). In the $0.8 < z < 1.0$ region, coercivity is considerably higher than for the rest of the alloys, whereas magnetostriction constant drops with the increase of z, and the mean grain size is of the order of 20 nm. In order to find out the reason for this unexpected behaviour, XRD results were included into the analysis. From Fig. 3 it is easy to find out that for $0.8 \leq z \leq 0.9$, there are two crystalline phases: bcc-Fe(Co) and fcc-Co(Fe). The latter one exhibits a higher magnetocrystalline anisotropy, so the scarcely distributed fcc-Co(Fe) grains act as the anchors of the domain wall movement, although they meet the requirement of ultrafine structure. This phenomenon is similar to the early stage of nanocrystalline alloys crystallisation: when there are only a few grains, their magnetocrystalline anisotropy can not be averaged out in the volume defined by the domain wall width, so each grain is a single crystal with considerable magnetocrystalline anisotropy. This phenomenon was first observed and described in [8]. In the case of HITPERM alloys with $0.8 \leq z \leq 0.9$, the mixture of bcc-Fe(Co) grains embedded in an amorphous matrix is magnetically soft (see e.g. the $z = 0.5$ or 0.6 alloys), but its magnetic softness is spoiled by separate isles of fcc-Co(Fe) grains. Therefore, high cobalt content is clearly detrimental to the desired properties of the alloys designed for cores of transformers, chokes etc.

In the case of applications such as sensors and actuators, magnetic softness is not the principal property – here, high magnetostriction coefficient should be sought, although magnetic softness is not a disadvantage. As mentioned earlier, high values of λ_s in HITPERMs are found when z is of about 0.5, which confirms the results reported earlier [9]. The increase of Co content above $z = 0.5$ will result in the reduction of magnetostrictive effect, and in the worsening of coercivity, which, eventually, impairs the functional properties of a sensor or actuator. What is also important, the HITPERM alloys are used predominantly in the high temperature environment, which requires that Co content should be high enough to prevent loss of magnetic softness and induction in the working environment. These factors imply that HITPERM-type alloys for high temperature applications for sensors and actuators should contain equal amounts of iron and cobalt.

Summary

In this work, analysis of the influence of magnetocrystalline and magnetoelastic anisotropies on magnetic and magnetoelastic behaviour was analysed on the example of nanocrystalline (Fe,Co)-(Hf,Zr)-Cu-B alloys. It was found that cobalt content has the major impact on both types of anisotropy, and in the case of HITPERM-type alloys the magnetoelastic anisotropy plays the main role in determination of useful properties of nanocrystalline, magnetically soft alloys. From the application viewpoint, for magnetically soft components, Co content should be as low as possible, and careful control of structure is required; for sensors and actuators, cobalt content should be equal to iron content, and careful control of structure is not as important, as in the case of magnetically soft applications.

Acknowledgment

This work was supported, as a targeted grant No. PBZ-KBN-115/T08/2004 in the years 2005-2008, from the funds for scientific research of Polish Ministry of Science and Higher Education.

References

[1] Y. Yoshizawa, S. Oguma and K. Yamauchi: J. Appl. Phys. Vol. 64 (1988), p 6044.

[2] K. Suzuki, N. Kataoka, A. Inoue, A. Makino and T. Masumoto: Mater. Trans. JIM Vol. 31 (1990), p. 743.

[3] M.A. Willard, D.E. Laughlin, M.E. McHenry, D. Thoma, K. Sickafus, J.O. Cross and V.G. Harris, J. Appl. Phys. Vol. 84 (1998), p. 6773.

[4] G. Herzer: IEEE Trans. Magn. Vol. 25 (1989), p. 3327.

[5] T. Kulik, H.T. Savage and A. Hernando: J. Appl. Phys. Vol. 73 (1993), p. 6855.

[6] G. Vlasák: J. Magn. Magn. Mater. Vol. 215-216 (2000), p. 479.

[7] J. Ferenc, M. Kowalczyk and T. Kulik: unpublished results.

[8] T. Kulik, A. Hernando and M. Vazquez: J. Magn. Magn. Mater. Vol. 133 (1994), p. 310.

[9] M. Müller, H. Grahl, N. Mattern, U. Kühn and B. Schnell: J. Magn. Magn. Mater. Vol. 160 (1996), p. 284.

Solid State Phenomena Vol. 154 (2009) pp 175-179
online at http://www.scientific.net

Magnetic field analysis for magnetron sputtering apparatus for accurate composition control

Yutaka Sakurai[1, a], Ryo Nakajima[2] and Hiroko Nakamura[3]

[1]Toyama National College of Technology, 13 Hongo, Toyama 939-8630, Japan

[2]YKK Corporation, 200 Yoshida, Toyama 938-8601, Japan

[3]Hokuriku Electric Power Company, 15-1 Ushijima, Toyama 939-8686, Japan

[a]ysakurai@toyama-nct.ac.jp

Keywords: Magnetron Sputtering, Thin Film, Composition Control, Magnetic Field Analysis, Finite Element Method

Abstract. Authors use magnetron sputtering technique for controlling the film composition by modifying the magnetic field with an external solenoid in addition to the magnetic field with a permanent magnet on back of composite target. It is necessary to understand the contribution of the solenoid quantitatively for the effective application of this technique. The magnetic field changes by the solenoid current on the target were calculated by the finite element method (FEM), and compared with the film composition. As the solenoid current increases, magnetic tunnel region on the target (correspond with the well sputtered region by the confined plasma) moves to the centre of the target. The behaviour corresponds with the actually formed film composition. The calculated results also give an information to design the composite target and the correction value for using the already eroded target.

Introduction

We have been forming binary alloy films by using magnetron sputtering method and additionally applying external solenoid current[1], [2]. The method enables us to control composition of NiTi[3] and/or TiW thin films easily by changing the solenoid current. To obtain more accurate composition control, however, errors of referred and controlled film composition have not been neglected. A significant reason of the error may be a surface field changes by the target consumption. Therefore, we performed the analysis to obtain the field distribution and the changes by internal and external conditions. FEM (Finite Element Method) allows the calculation taking into account shapes, materials parameters, currents, and permanet magnets, so it was a suitable method for this purpose.

The calculated results were compared with film composition formed actually, and the controllability of film composition was well interpreted by assuming the position that vertical field equals to zero (B_y=0) as a well sputtered region. The film composition error by consuming the target was analyzed as an effect coming from the field changes, and the correction values of the current were obtained.

Objective and analysis

Sputtering apparatus Fig. 1 shows schematically illustrated cross-sectional view of our objective sputtering apparatus. A solenoid conductor is additionally employed to the conventional DC magnetron sputtering system with a composite target for forming binary alloy films. Fig. 2 shows the more detailed structure of the target and the magnetic circuit.

Objective region of apparatus for FEM and sputtering origin Fig. 3 shows schematic illustrations of the sputtering apparatus and objective region for FEM (left) and the sputtering situation (right). The FEM and magnetic field analysis will be explained at the next section, a significant reason of the magnetic field analysis is given here.

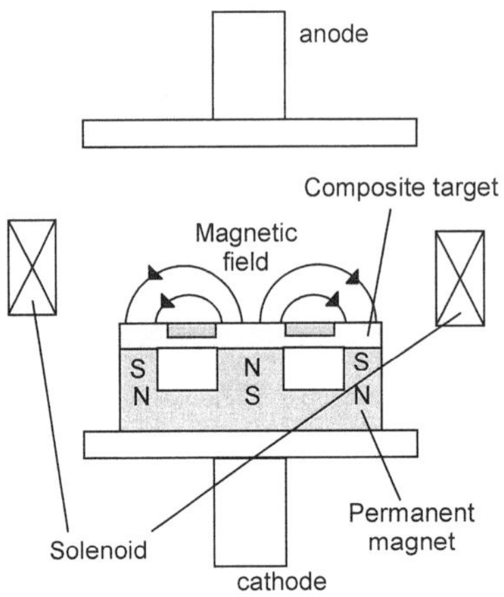

Fig. 1. Schematic view of the objective sputtering apparatus.

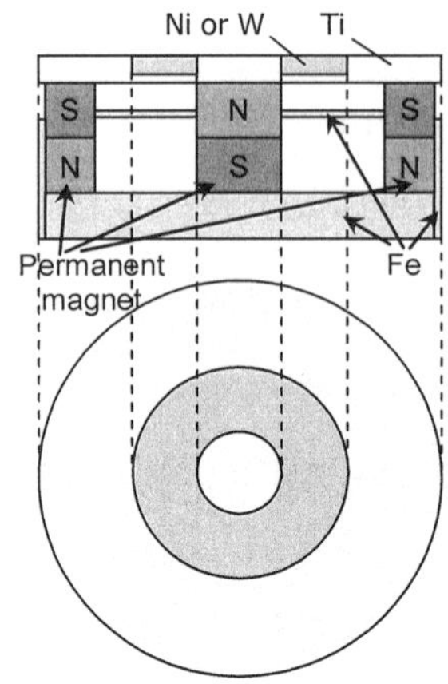

Fig. 2. Schematics of the structure of the composite target and the back side magnetic circuit (upper: cross-sectional side view, lower: top view).

As shown in the right picture of Fig. 3, magnetic tunnel where plasma is confined, is conventionally formed by the magnets on back of the target. In our apparatus, the magnetic tunnel is moved by the field of solenoid current from outside, and the film composition can be easily controlled. Obtaining the position of the magnetic tunnel is significant to accurate film composition control since the position corresponds to a well sputtered region and is shifted with target consumption. This means that different film compositions are obtained at the same current condition. If experimental data of the solenoid currents vs. film composition is obtained in a condition by using a non-eroded target, the data can not be applied in the other conditions by using an eroded target. Therefore, to realize an accurate film composition control experimentally, a lot of data acquisition is needed. Fortunately, magnetic field distribution on the target might be a dominant factor in the film composition control, so a fine condition control might be possible by using the non-eroded target data and the current correction from the field calculation.

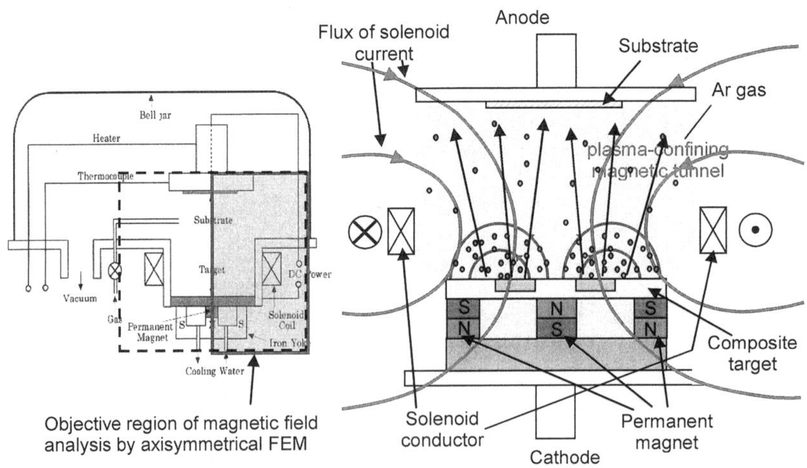

Fig. 3. Schematic illustration of the sputtering apparatus and objective region for FEM (left) and the sputtering situation (right).

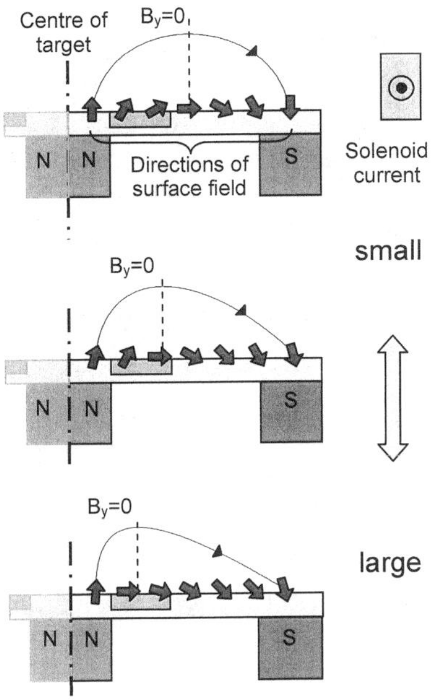

Fig. 4 Illustration of moving B_y=0 region by changing solenoid current. The cross-sectional side views of the sputtering target and the composite fields are shown. The situation of smaller solenoid current one is shown in upper figure, and larger one in lower figure.

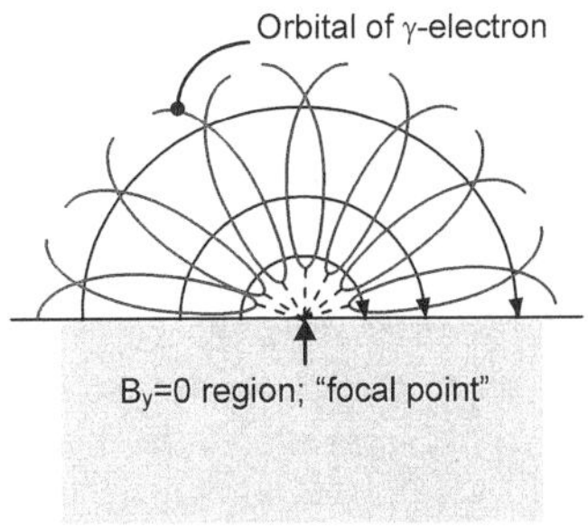

Fig. 5 Cross-section of "magnetic tunnel" on the target. The region that vertical component of the field is equal to zero (B_y=0) is assumed as the most sputtered region.

Fig. 4 schematically shows the moving behaviour of the magnetic tunnel. Top of the magnetic tunnel is moving toward the centre with current increases in these directions of the magnet and the current.

Fig. 5 depicts a focal point of γ-electron confined around the magnetic tunnel shaped field. We assumed that the focal point is a position on the target that the vertical component of magnetic field equals to zero (B_y=0) for convenience to evaluate. We consider that the focal point is the most sputtered region on the target.

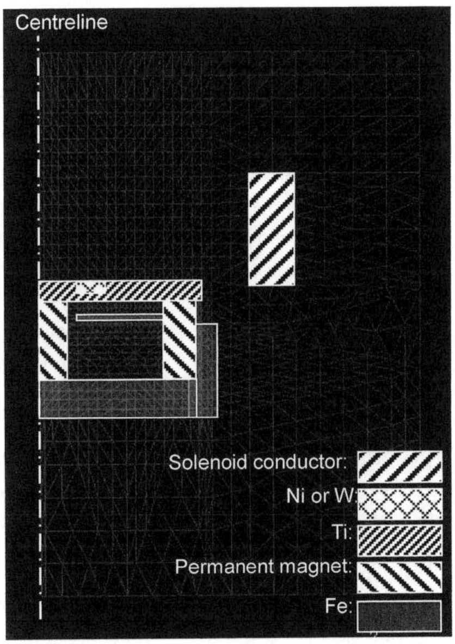

Fig. 6 Mesh for the FEM calculation. The number of nodes is 1216, and elements 2323.

Magnetic field analysis Magnetic field calculation was performed by using the FEM (Finite Element Method). The 1-st order triangular element was applied for the axisymmetrical cross-section of the apparatus. Fig. 6 shows the mesh for the analysis region divided by the triangular elements and materials taken into account on the mesh. The material constants were adequately given to the calculation. Magnetic relative permeability of ferromagnetic Ni and Fe were given as 50, and the magnetization value of permanent magnets was decided by doing trial and error as agreed with measured target surface field. Boundary conditions were free except for the left side corresponding to the central axis of axisymmetrical geometry. Target erosion was assumed as a uniformly proceeded case for a convenience of analysis.

Result and discussion

Flux distribution (without erosion) Fig. 7 shows an example of flux distribution in case of the Ti/W composite target and the solenoid current is zero. We can see the movement of the magnetic tunnel with increasing the solenoid current, actually, the movement is connected with the moving B_y=0 point assumed as above.

Ti/W composite target Fig. 8 (a) also shows experimental TiW film composition vs. solenoid current. The composite target structure is schematically shown in the figure. W-content gradually increases with solenoid current increases, film composition changes per applied solenoid current is approximately 5 at%/A. Fig. 8 (b) shows the vertical component of the field as a function of distance from the target centre. B_y=0 region moves to W-region with solenoid current increases, and the moving halts in the place. It means that a limitation of W-content of the film is exist in this composite target. In order to obtain the more W-rich film, W-region should be more extended.

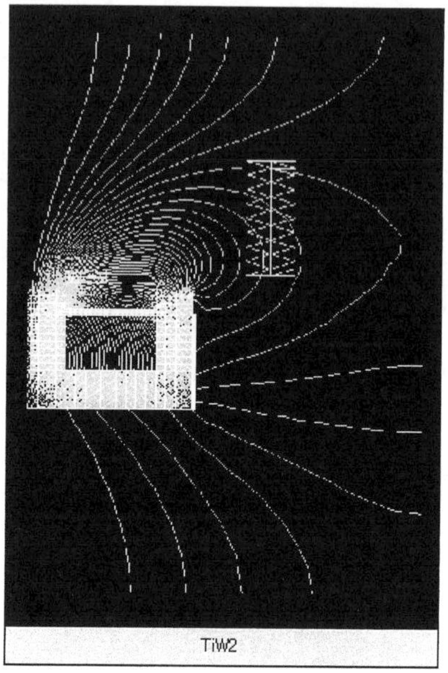

Fig. 7 An example of calculated field distribution in the objective region. The calculation was performed any times corresponding with various solenoid currents.

Ti/Ni composite target Fig. 9 (a) also shows experimental TiNi film composition vs. solenoid current. Ni-content linearly increases with solenoid current increases, film composition changes per applied solenoid current is approximately 10 at%/A. Fig. 9 (b) also shows the vertical component of the field as a function of distance from the target centre. B_y=0 region approaches to Ni-region with solenoid current increases. The second B_y=0 region occurs at larger current values than 1A. Appearance of the second B_y=0 region might be a existence of the ferromagnetic material of Ni on the target surface. This means that Ni-content is successfully controlled, and the target design is valid, especially forming 50-50 film corresponding to a shape memory alloy film.

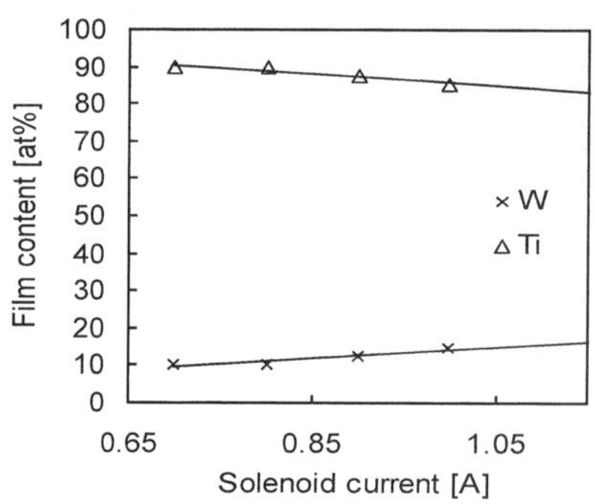

(a) Experimental film content vs. solenoid current.

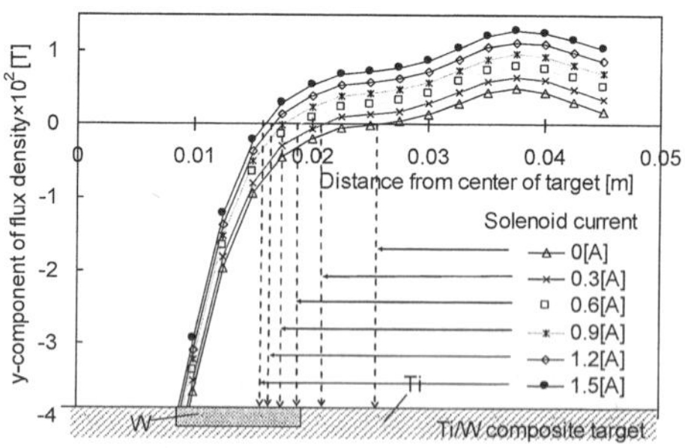

(b) Calculated vertical field distributions on the target as a parameter of solenoid current. B_y=0 points of various solenoid currents on the composite target are shown by the dotted lines on the figure.

Fig. 8. Experimental result (a) Experimenta land calculated field distribution on the target (b) in case of Ti/W composite target.

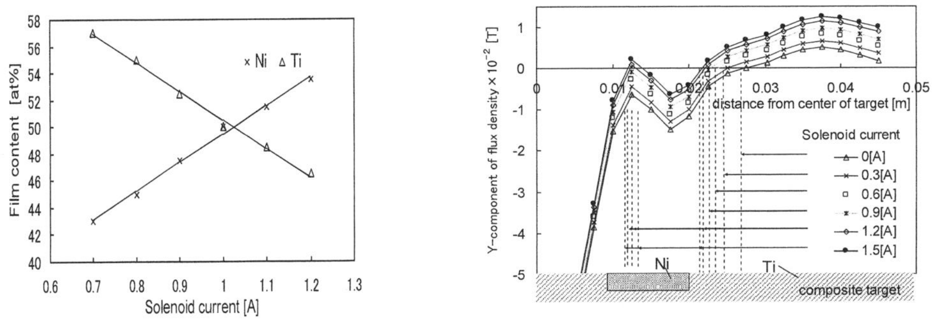

(a) Experimental film content vs. solenoid current.

(b) Calculated vertical field distributions on the target as a parameter of solenoid current. B_y=0 points of various solenoid currents on the composite target are shown by the dotted lines on the figure.

Fig. 9. Experimental result (a) and calculated field distribution on the target (b) in case of Ti/Ni composite target.

In case of eroded target Fig. 10 shows solenoid current vs. distance of B_y=0 region from target centre as a parameter of erosion depth. Solenoid current should be smaller to obtain the same condition as non-eroded target using eroded target. The current correction value corresponding to the target erosion is obtained by reading current at same B_y=0 position on the figure.

Conclusion and future work

The relationships of magnetic field distribution vs. solenoid current in case of Ti/W and Ti/Ni composite target were calculated by using FEM, and compared with the film composition experimentally acquired by the apparatus, and the followings were obtained.

(1) The calculation results were well organized with experimental film composition, and effective to design the composite target.

(2) Current correction values for using the eroded target were obtained.

In future work, re-designing composite target, and form the films to reconfirm the correspondence of geometrical structure of the target and film composition. Moreover, in this evaluation, static magnetic field analysis were applied, and the well sputtered region was assumed as the B_y=0 point, however, dynamic coupling simulation of electromagnetic and particle movement might be needed by more detailed evaluation.

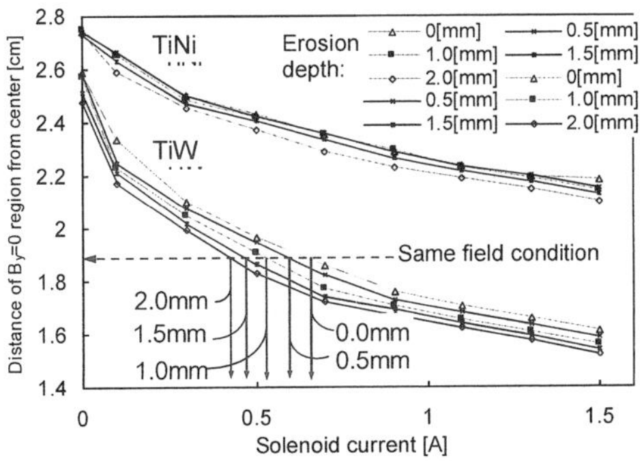

Fig. 10 Distance of B_y=0 region from target centre vs. solenoid current as a parameter of erosion depth. Solenoid current should be smaller to obtain the same condition as non-eroded target using eroded target.

References

[1] F. Takeda and T. Nakajima: Thin Solid Films, 336, 68 (1998)

[2] F. Takeda, T. Yamazaki, and T. Nakajima: Jpn. J. Appl. Phys., 39, 5992 (2000)

[3] Y. Sakurai and F. Takeda: Abstract of E-MRS Fall Meeting (2005)

Solid State Phenomena Vol. 154 (2009) pp 181-186
online at http://www.scientific.net

3D magnetovision scanner as a tool for investigation of magnetomechanical principles

J. Kaleta, D. Lewandowski, P. Wiewiórski

Wroclaw University of Technology, Institute of Materials Science and Applied Mechanics, Smoluchowskiego 25, Wrocław 50-370, Poland

jerzy.kaleta@pwr.wroc.pl, daniel.lewandowski@pwr.wroc.pl, przemyslaw.wiewiorski@pwr.wroc.pl

Keywords: magnetomechanical effects, NDE, magnetovision, magnetovision camera

Abstract. The study was aimed at designing a system for measuring the distribution of magnetic field around different magnetic objects including Smart Magnetic Materials. A new type of 3D camera for monitoring the magnetic field intensity was constructed. The measurement principle is based on internal magnetics properties of materials and the reverse magnetostriction effect (also called the Villari effect). No external magnetizing field is assumed; the entire magnetic effect is due to magnetomechanical principles. A new generation of Honeywell magnetosensors were applied to measure field intensity in 3D. Small size of measurement area (1.5 mm^3) allows quasi-local measurement of magnetic field. In the measurement head also Hall probes were alternately used. The aim of this stage of research was to construct the set for measurement of strength of a very weak magnetic field (10 A/m) around the magnetic objects. In scanner construction three axes for displacement in directions consistent with sensor axis were applied, which permits measuring magnetic field vector in geometrical coordination. Specialized software for data acquisition, processing and visualization of magnetic field vector has been written. In preliminary parts of the work magnetic scanner system allows determination of correspondence between mechanical and magnetic quantities. Main applications for this type of system are: reverse magnetostriction and magnetostriction in smart magnetic materials and composites, martensitic transformation induces plastic strain in shape memory alloys, NDT investigation, identification of local plastic deformation and texture of ferromagnetic materials, magnetic polygraphy and others. Measuring system may be used both as an entirely autonomous system as well as an integrated one, also through joint control with a typical mechanical testing systems for static and fatigue tests.

Introduction

Various devices for optical visualization are commonly used in techniques (optoelectronic cameras) and for thermal visualization (thermal infrared imagining cameras). In recent years there have been attempts at visualization of the magnetic field by means of the so-called scanners and magnetovision cameras. These devices make it possible to work out a new method from groups NDT and NDE called magnetovision, which has a number of very important applications in technology and medicine. All ferromagnetic materials can be an object of experiments. Most interesting area of applications of magnetovision is connected with making use of the inverted magnetostriction (also referred to as Villari effect and examining Smart magnetic materials [1,6]. The present scientific and engineering aim of such experiments is also identification of physical models of Villari effect, which would allow determining the deformation field on the basis of vector's components of the magnetic field intensity. The reasons presented above provided the present authors with motivation for creating the third generation scanner - magnetovision camera featuring the most up-to-date achievements in the whole range of sensors, software and mechanics. For over a decade in the team the present writers recruit from, a few generations of cameras and scanners have been developed, which ensure a magnetic image of materials exposed to various kinds of treatment, mechanic load specters and phase transformations. Earlier works present the basics of new diagnostic technique called magnetovision and point out the possibility of developing measuring apparatus (magnetovision). The earlier works also present metrological properties of the effect [2,3,5,6], the possibility of application in fatigue process tests, in tests of breaking and texture and in athermal martensitic transformations. Not only can the system presented below be easily put to the uses described earlier, it has even a wider range of applications.

The setup of the magnetic scanner

The scheme of the system called Magscanner is presented in Figure 1, and Figure 2 presents its front view. The system consists of the following elements:

- mobile system of control and operation of the experiment consisting of the IBM computer T30 with a docking station, and service of full-size PCI cards,
- advanced control system ADDI-DATA APCI8001, mainly used in robotics
- scanner XYZ, advanced device adapted also for work with mechanical testing machines MTS. Guaranteed resolution of the measurement head scanning movement – 2160 DPI. Working area of scanning: 410mm x 180mm x 200mmm,

A)

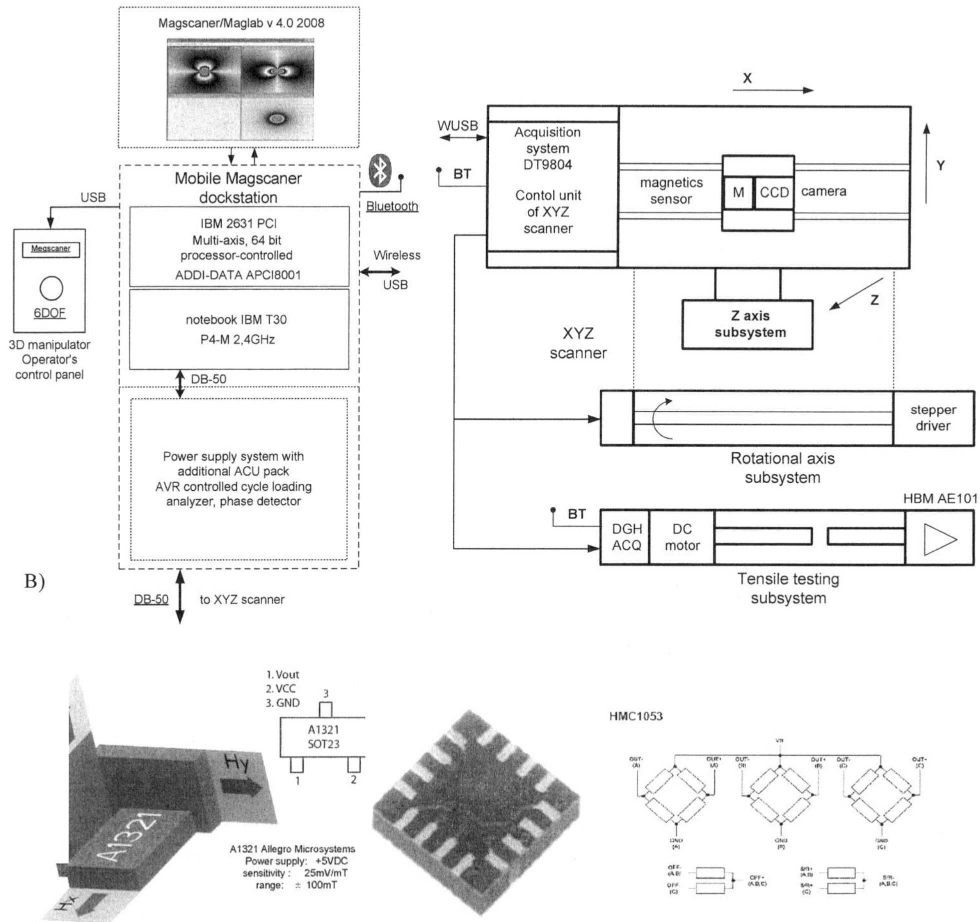

Figure 1. Practical realization of the mobile magnetovision system adapted for complex tasks of determining of the magnetic field distribution in different applications A) applied measurement heads and their structure B).

- measurement head in the shape of triaxial measurement head based on Honeywell HMC105 magnetoresistors. These heads are easy to replace and measurements can also be made with Philips KMZ magnetoresistors as well as Hall elements made by Allegro Microsystems. A probe with sensors moves along the examined object at a constant distance (usually 0.2 – 0.5 mm) from it,
- data acquisition system Data Translation (DT9804) with the so-called trigger for measurement and defining of the carriage positions,
- software dedicated to Magscanner/Maglab operating the whole of the system and ensuring results visualization. This package allows export of results to parameter systems CAD and visualization of the magnetic field vector 3D space,
- rotary axis designed for analysis of the magnetic field distribution around the axisymmetric elements – cylinders and pipes. Main emphasis was laid on the possibility of quick tests done with a view to the scanner's use in industry,
- breaker – a device enabling precise measurements of the magnetic field and mechanical values for materials in the shape of foil in tests of static breaking. Resolution of scanning movement of the beam equals 0.25 micrometers. Thanks to the use of the HBM AE101 class 0.1% amplifier it was possible to attain a large resolution of the force measurement in the range up to 500N.

Measuring parameters of the scanner and measurement head

Main parameters of the signal acquisition, their processing results visualization have been set out below:

- range of the magnetic field measurement - depending on head – up to 100mT (hallotrons)
- scanning speed up to 50 000 points /sec,
- scanning resolution – max. 2160 DPI (0.02mm)
- max. magnitude of XY map of magnetic field 20 million points (20000x1000),
- card sampling frequency DT 9804 – 100 kHz, at 16-bit resolution.

Fig. 1 presents the scheme of a measurement head with Hall elements. An integrated probe was made by sticking three measurement elements together (for Hx, Hy, Hz components respectively), each of which has its own independent conditioning system. Concentration on a small surface of a complete measurement setup of the magnetic field based on Hall effect allows quasi-point measurements, essential for the cylinder surface being analyzed. The measurement resolution attained equaled 100 μT, in the linear range of the sensor ± 100mT.

Figure 2. View of the Magscanner (ver. 2007) system, in which control-measurement part of the XY scanner (A), breaker (B), XY scanner (C) and mobile position of the magnetovision camera (D)

Selected examples of the scanner's applications

Examples of the scanner's applications and earlier versions of the magnetovision camera were presented in our earlier papers. Three cases have been chosen to illustrate its uses. In the first two magnetoresistors were used in order to perform measurements, and in the materials under examination occurred Hall effect. Magnetic field changes were very small of the order of earth background magnetic field. Figure 3 shows the results of the Villari effect measurement for Kirsh specimen exposed to cyclic load.

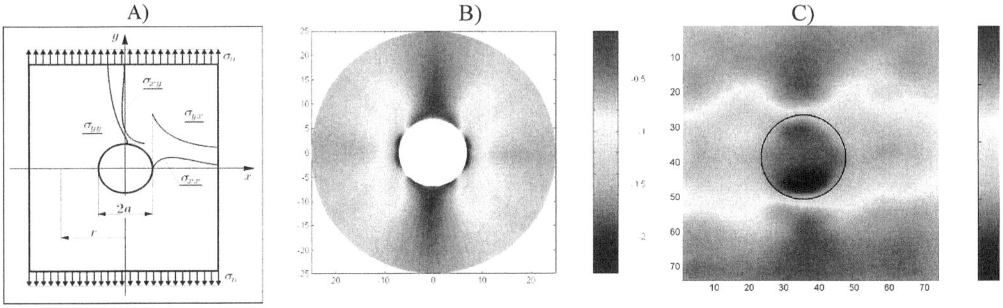

Figure 3. Kirsch specimen and stress state (a model) A), the map of a deviator of stress state along load axis for Kirsch specimen (a model) B), The magnetovision image of the Kirsch specimen under load C) [2,4].

The second group of measurements was performed so as to check the usefulness of the magnetovision system for determining the magnetic field around objects submitted to technological procedures with a view to checking their correctness (e.g. cutting, laser ablation, electro-discharge drilling, micro-layer plotting, quality of magnetic printing, etc.) Figure 4 contains selected examples.

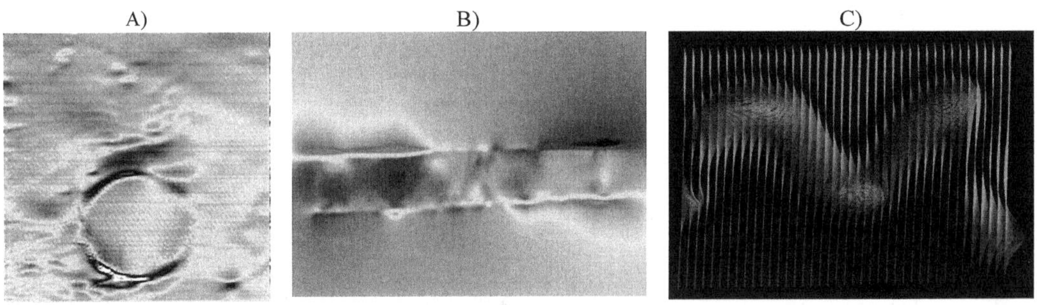

Figure 4. Examples of the use of the magnetic field stream disturbance for detection of changes on the ferromagnetic's surface A) - metal plate with drilled hole, B) - steel strip horizontally (changes of magentic field shows deformation and cutting edges), C) - vector field observed over thin cupper wire (current flow through it).

In the third successive case (Fig. 5 and Fig. 6) composite magnetic rods made of strontium ferrite were the object of our experiments. The cylinder's diameter equaled 9.60 mm and its working length – 220 mm. In order to verify the experimental method we used cylinders locally demagnetized but without mechanical defects and cylinders with introduced 'programmed' mechanical defects in form of drilled holes, defects and an induced crack. Reference cylinders had neither mechanical defects nor demagnetized areas. To perform the measurement effectively we had to rearm the measurement head by having a three-axis Hall probe mounted on it. The probe moved at a constant distance (0.5 mm) from the cylinder.

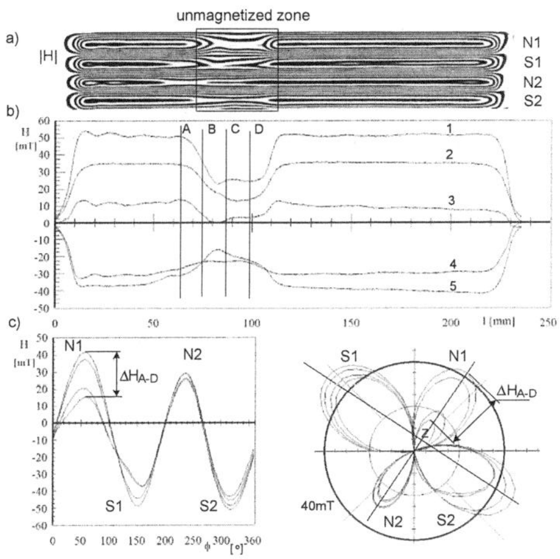

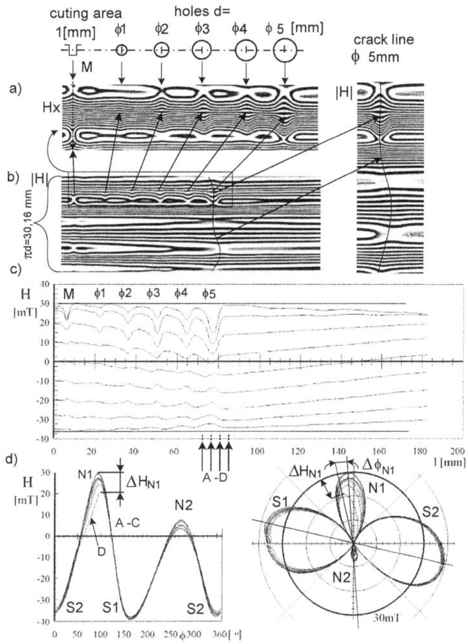

Figure. 5 Results of tests for a cylinder subjected to a local influence of external magnetic field. a) magnetic field map b) distributions of magnetic field [H] for successive longitudinal cuts c) d) distributions of changes [H](l (mm), φ (angle)) for marked transverse cross sections and (on the right side) their image in polar coordinates.

Figure 6. Results of tests for a cylinder subjected to local mechanical defects. a) The manner and place of applying mechanical defects, and magnetic field map [H] with marked crack place and its enlargement [next to] distributions of changes [H](l (mm), φ (angle)) for transverse cuts lying in the area of the opening 5mm and a detected crack.

Conclusions

1. A new generation scanner (magnetovision camera) with the replaceable head containing magnetoresistors and Hall probes has been developed.
2. The scanner can find application as one of the NDE methods for testing the quality of diverse technological processes and for identification of cross models of magnetomechanical field.
3. The scanner system can be used as an entirely autonomic one and as commonly controlled together with a typical material testing systems for static and fatigue experiments

Acknowledgment

This research was carried out partially as part of Commissioned Project PBZ-KBN- 115/T08/2004 financed by the Ministry of Education and Science, the State Committee for Scientific Research.

References

[1] J. Kaleta, *Magnetomechanical Effect in Fatigue Investigations of Ferromagnetics*; Proc. of EUROMAT-2000, Tours, 7-9 November 2000, France, Vol.2, s.1149-2254; 2000; Tours.

[2] J. Kaleta, P. Wiewiórski , *Magnetovision as a Tool for Investigation of Fatigue Process of Ferromagnetics*; Proc. of Fatigue 2001, Sao-Paulo 12-14 December; 7; 2001; Sao Paulo.

[3] J. Kaleta , P. Wiewiórski, W. Wiśniewski , *Badanie tekstury blach ferromagnetycznych z wykorzystaniem efektu Villariego*, Sympozjum Mechaniki Zniszczenia Materiałów i Konstrukcji, 2003, Augustów.

[4] J. Kaleta, R. Górecki, *Metoda wizualizacji pól krzyżowych w tarczach obciążonych cyklicznie. Zmęczenie i mechanika pękania*; Materiały XIX Sympozjum Zmęczenia i Mechanika Pękania; 2002; Bydgoszcz.

[5] J. Kaleta, P. Wiewiórski, W. Wiśniewski, *Magnetyczna pamięć historii odkształcania materiału ferromagnetycznego*; IX Krajowa Konferencja Mechaniki Pękania. Kielce-Cedzyna, 14-17 września 2003. Kielce: Wydaw. PŚwiętokrz. 2003 s. 221-228.

[6] J. Kaleta, P. Wiewiórski: *Detekcja defektów w magnetycznych prętach kompozytowych z użyciem skanera o dużej rozdzielczości*. Acta Mechanica et Automatica. 2007 vol. 1, nr 1, s. 39-44.

Solid State Phenomena Vol. 154 (2009) pp 187-194
online at http://www.scientific.net

Smart Technologies for Adaptive Impact Absorption

J. Holnicki-Szulc[1, a], C. Graczykowski[1, b], G. Mikułowski[1, c], A. Mróz[1, d], P. Pawłowski[1, e],

[1]Institute of Fundamental Technological Research, Polish Academy of Sciences, Świętokrzyska 21, 00-049 Warsaw, Poland

[a]holnicki@ippt.gov.pl, [b]cgraczyk@ippt.gov.pl, [c]gmikulow@ippt.gov.pl, [d]amroz@ippt.gov.pl, [e]ppawl@ippt.gov.pl

Keywords: Adaptive Impact Absorption, Adaptive structures, Optimal control.

Abstract. The article presents a review of recent research carried out in the Department of Intelligent Technologies of Institute of Fundamental Technological Research, dedicated to application of systems for adaptive impact absorption to adaptive aircraft landing gears, novel concept of protective MFM structures, flow-control based airbags, maritime applications of inflatable structures, and development of adaptive wind turbine blade – hub connections.

Introduction

Increasing demand for safety becomes nowadays a clearly visible trend. One of main areas of widespread research is design of systems protecting against heavy dynamic loads such as low and medium velocity impacts or environmental loadings. Commonly applied passive systems are typically designed for a specified load scenario and therefore their performance over a wide range of expected loads is limited. This shortcoming can be significantly reduced by application of systems of Adaptive Imapct Absorption (AIA). The general concept of AIA (e.g. [1], [2]) refers to the application of intelligent technologies, which focus on adaptation of energy absorbing structures to extreme overloading by impact identification and application of controllable dissipaters based on magneto-rheological fluids, piezo- or magnetostrictive- actuated devices.
The article presents a review of recent research carried out in the Department of Intelligent Technologies of Institute of Fundamental Technological Research, dedicated to application of systems for adaptive impact absorption to aircraft landing gears, a novel concept of protective MFM structures, flow-control based airbags, maritime applications of inflatable structures and development of adaptive wind turbine blade – hub connections.

Adaptive Multifolding Structures

The model multifolding structure (MFM c.f.[3]) is composed of truss-like elements arranged into a special pattern, depicted in Fig. 1. The elements are equipped with micro-devices (so-called "micro-fuses") which control the axial force in each element. The stress thresholds, triggering the micro-fuses, are uniformly distributed along horizontal layers but may change in the vertical direction. Therefore, during deformation, layers with the lowest stress values collapse first. The repetitive use of elements (multifolding effect, e.g. element 2 in Fig. 2-left) provides the synergistic effect in the process of energy dissipation. Two selected deformation modes of the most basic MFM system are depicted in Fig. 2.
The initial distribution of the control thresholds plays a crucial role in the adjustment of the stiffness characteristics of the MFM to the impact loading. Different initial distributions of those parameters will result in different folding sequences and therefore will change the capability of energy absorption and the level of acceleration during the impact.
The concept of multifolding was verified experimentally. The tests aimed at implementing models and control strategies applied in the theoretical and numerical studies.

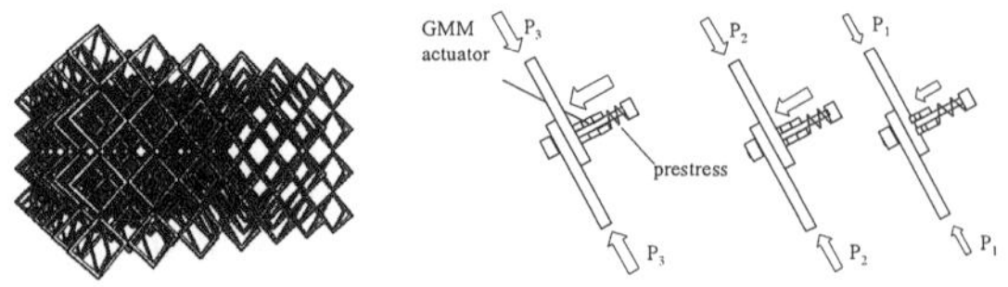

Figure 1. The layout of Multifolding Structure (left); Concept of "structural fuse" (right)

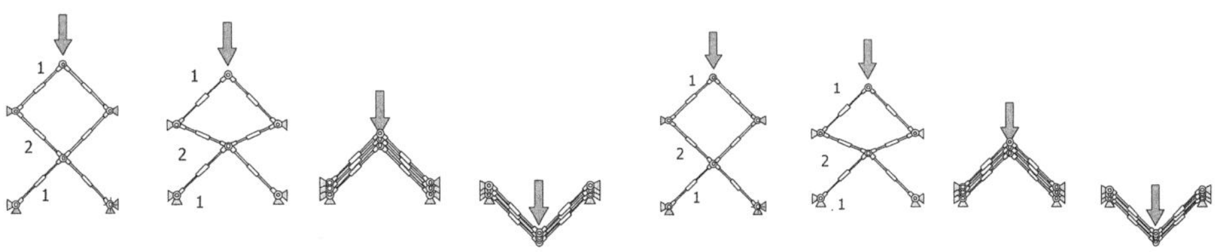

Figure 2. Selected folding patterns of basic MFM structure: mode 2 (left) and mode 1 (right)

The experimental set-up represents the most basic MFM substructure, composed of six controlled elements, which are magnetorheological fluid dampers. Nodes of the structure are guided by slide bearings (providing only one axis of movement) and are equipped with force and displacement sensors. The design of the set-up gives possibility of realising different folding sequences and control algorithms of the MFM structure.

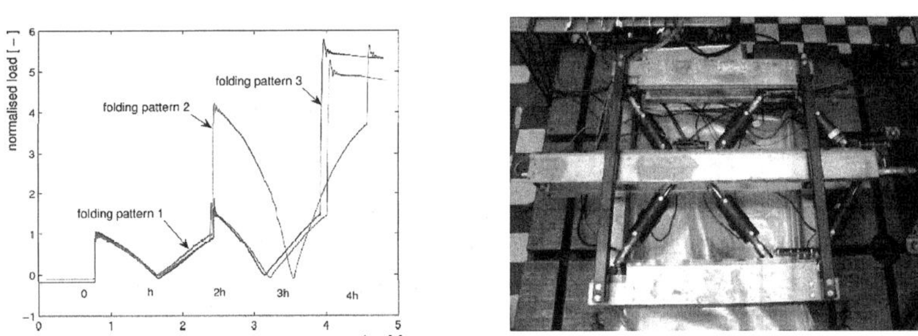

Figure 3. Experimental load curves for different folding patterns of basic MFM structure (left); View of experimental set-up (right)

Adaptive flow control based airbags

Airbag systems are commonly used in automotive industry to provide safety of the occupants during collisions since 1980s. Despite many years of development and improvement car airbags remain passive systems where only initial inflation is adjusted to actual impact scenario. After airbag deployment gas is released by fabric leakage only and no precise control of internal pressure is performed. This indicates that airbag behavior is still not optimal and can be significantly improved by introducing controllable gas exhaust.

Adaptive flow control based airbags' are deformable inflatable cushions made of rubber or fabric equipped with fast inflators and additionally with controllable high speed and stroke valves. The performance of the adaptive airbags is based on three following stages: impact detection and identification; appropriate initial inflation; active adjustment of pressure executed by controlled gas release.

The initial velocity and direction of the impacting object can be measured e.g. by ultrasonic velocity sensors. Its mass and initial kinetic energy can be recognized during the initial stage of collision by using accelerometers and pressure sensors and by applying one of the impact identification procedures described in Ref.[1].
The main advantage of *adaptive* inflatable structure over *passive* one is that in adaptive structure gas exhaust can be controlled and adjusted to actual impact scenario. Development of optimal pressure release strategy is the main challenge related to adaptive inflatable structures. The objective of applied control is to protect the impacting (or impacted) object by minimizing its accelerations $a(t)$, internal forces $\sigma(t)$ or rebound velocity V_R
Identified mass and velocity are utilized for development of optimal control strategy for inflation and pressure release. Controlled gas exhaust can be executed by opening controllable High Performance Valves (HPV) based on multifolding microstructures or thermically activated membranes. Active pressure release allows to adjust global compliance of the pneumatic structure in subsequent stages of impact and to prevent excessive accelerations and forces in the system. Moreover it helps to control dissipation of the energy and to avoid hitting object rebound.
Numerical analysis of inflatable structure subjected to an impact load requires considering the interaction between its walls and the fluid enclosed inside. Applied external load causes large deformation of the structure and change of the capacity and pressure of the fluid. Pressure exerted by the fluid affects, in turn, the deformation of the solid wall and its internal forces.

Maritime applications of inflatable structures

Flow control based airbags can be effectively utilized to mitigate open sea collisions, cf. Ref.[4]. The inflatable structure that will be used for protecting offshore wind turbine against impacts of small ships is torus-shaped and surrounds the tower at the water level. The walls of the pneumatic structure can be made of rubber reinforced by steel rods or any other material which provides high durability and allows large deformations during ship impact. The dimensions of inflatable structure are limited to 2-3 meters in height and 1m in width due to requirements of fast inflation and pressure release.
To obtain better adaptation to various impact scenarios, the inflatable structure is divided into several separate air chambers located around the tower, cf. Fig. 4 a,b. Controllable valves enable flow of the gas from each chamber of the torus structure to environment and between adjacent chambers.

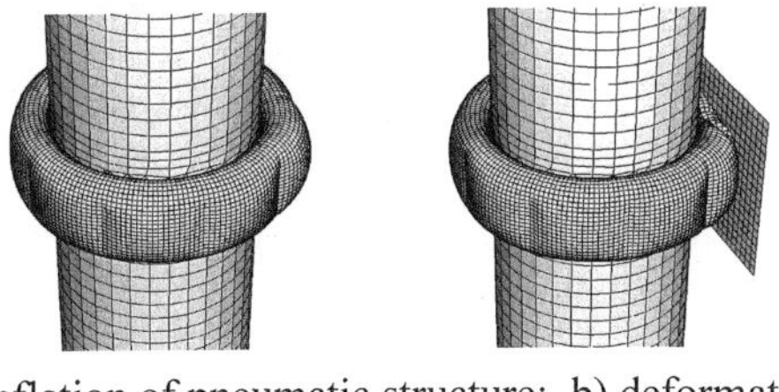

Figure 4: a) inflation of pneumatic structure; b) deformation during collision

The purpose of applying pneumatic structure is to mitigate the response of both the ship and the wind turbine tower. In particular, the inflatable structure helps to minimize ship deceleration, avoid ship rebound, decrease stresses arising at the location of the collision and mitigate tower vibrations. In case of active acceleration minimisation appropriate chamber valve opening (represented by flow coefficient) is proportionally adjusted on several time intervals and ship acceleration is maintained on desired almost constant, possibly low level, cf. Fig. 5.

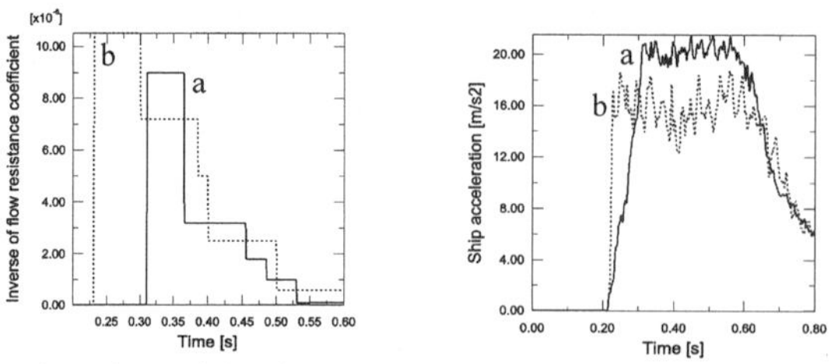

Figure 5: Active acceleration mitigation: a) pressure released during impact (continuous line), b) additional inflation at the beginning of impact (dashed line)

Adaptive airbags for helicopter emergency landing

Another applications of the proposed concept are adaptive external airbags for helicopter emergency landing. The system consists of four cylindrical cushions attached at outer side of helicopter undercarriage (cf. Fig. 6). The airbags are deployed and inflated just before touchdown by pyrotechnic inflators. During collision with the ground pressure is released by fabric leakage and by additional controllable high speed and stroke valves.

Initially the problem of helicopter stabilisation during landing was considered. For this purpose three dimensional model composed of stiff plate and four airbags was developed (Fig. 6a) and various landing directions and velocities were analysed (Fig. 6b). The control problem was to find initial airbags pressures and optimal (but fixed during landing) openings of each airbag valve for which landing scenario runs possibly smoothly i.e. the direct contact of the stiff plate and ground does not occur, the falling object does not bounce or rotate strongly and finally global measure of plate acceleration is minimised.

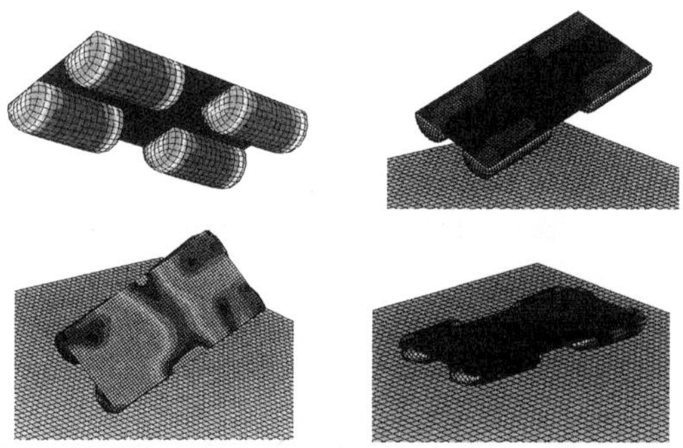

Figure 6: a) Considered model, b)landing scenario; c) non-optimal passive response with rear part rebound d) optimal uniform airbags compression

Another control problem was oriented towards minimization of stresses arising in helicopter undercarriage during landing. In numerical example the simplified two dimensional model of falling object composed of deformable beams and point mass was used (Fig. 7a). Three pressure adjustment strategies were applied (Fig. 7): i) optimisation of initial pressure only (continuous line), ii) optimisation of initial pressure and constant valve opening (dashed line):, iii) continuous control of valve opening to maintain precomputed optimal pressure pressure level (dotted line):

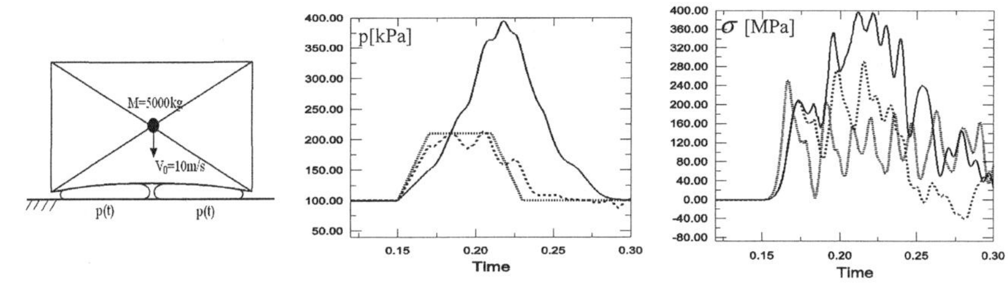

Figure 7: Active stress minimisation: a) considered model , b) applied pressure variation , c) resulting stresses at lower beam of the model

Adaptive pneumatic landing gear for Unmanned Air Vehicle (UAV) application

Landing gears are considered by the aeronautic designers as indispensable components of the aircraft. However, they are also aware that undercarriage disturb the aerodynamics and limit significantly the payload of the aircraft due to its own weight. Considering only these two mentioned constraints, it can be concluded that the optimal landing strut should have the minimal possible weight and the minimal possible volume.

The presented problems in design of the landing gears might be omitted via utilization of the concept of pneumatic adaptive impact absorber. Primarily, a design of gas shock absorbers for landing gars will let reduce the effective weight of the structure. Secondarily, the introduction of the smart technology allows execution of the gas migration management in the shock absorber and therefore to adapt the characteristics of the absorber to the actually recognized energy of the landing impact.

The idea of introducing active systems with capability of controlling the behaviour of landing gear struts was considered since the 1970s. Most of the concepts were based on the idea of influencing the shock absorber performance by regulating the internal fluid pressure over time. The initial concepts developed e.g. in NASA Langley Research Center, were followed by first experimental works in the 1980s, which indicated that utilised hydraulic control systems were not fast enough to perform an efficient control procedure during impact phase of the landing. More promising semi-active concept, which was based on real-time control of a piezo-valve adapted to landing gear, was the result of the research project ADLAND (FP6 2002 Aero1 - 502793).

Figure 8: Concept of adaptive pneumatic absorber

The proposed solution for adaptive landing gear is based on double chamber pneumatic cylinder equipped with active piezoelectric or magnetostrictive valve which controls flow of the gas between cylinder chambers, cf. Fig. 8. Proposed device can serve as a shock absorber dissipating UAV kinetic energy during landing and as suspension for taxiing and ground maneouvres. Active adaptation of the pneumatic absorber to actual landing scenario comprises the two following steps: initial identification of aerial vehicle kinetic energy based on ultrasound measurement of touchdown velocity; active control of the piezoelectric valve opening providing constant pneumatic force generated by the absorber and uniform energy dissipation during landing process.

Active control strategy is executed by control system where UAV mass and touchdown velocity are introduced as known input parameters. Actual level of pneumatic force is used as a reference signal for closed-loop feedback system.

The proposed adaptive device was modeled numerically in order to assess system effectiveness and optimal geometry, range of pressures and valve parameters to be applied in final design of UAV absorber.

The presented hardware concepts were implemented in a lab-scale experiment and verified on a drop test stand (cf. Fig 9a). The pneumatic cylinder was equipped with an adaptive valve characterised by response time equal 2 ms .

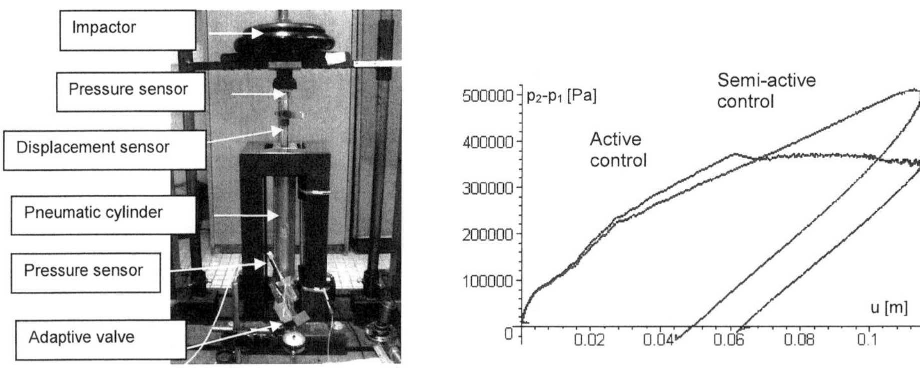

Figure 9: a) Experimental stand. b)Comparison of the pressure difference in the cylinder's chambers for semi-active and active control modes.

Successive tests conducted on the lab-scale stand were verification of the system's control strategies under impact loading. Two control strategies (semi-active and active) under the same conditions of impact were established. Comparison between the results obtained with the semi-active and the active control strategies (Fig. 9b) pronounced reduction of the maximal transferred load value by 30% in the case of the active control strategy.

Adaptive wind turbine blade – hub connection

In order to meet the EU goal for the wind energy production for year 2020 it is expected that the rate of the market growth will be increasing, and considering that one big wind turbine is more efficient that many small ones, it is expected that also the size of wind turbines has to be increasing. There are, however technological barriers on the way to up-scaling, such as the weight limit, tip speed limit or the blade root bending moment. The blade root bending due to extreme wind gusts causes the blade root bending stress to be a design limiting factor. Two possibilities to overcome this barrier are new composite materials development on one hand and new adaptive solutions on the other. The latter is the subject of presented work.

A semi-active adaptation technique was proposed basing on the following observation. Since the aerodynamic torsional moment forces a blade to turn to feather, it can be expected that, once the torsional connection of a blade is freed, it could increase the blade pitch angle thus reducing the

blade loads caused by a gust. Consequently the root bending and resulting stresses could be also mitigated.
For the purpose of assessing the effectiveness of the proposed solution a simple wind turbine numerical model has been built. The model consists of aerodynamic, structural an adaptation modules. All degrees of freedom that influence the aerodynamic forces are included in the model. Control procedures can be applied to any degree of freedom.
The adaptation of the blade – hub connection is summarized in terms of the pitch angle changes. After the gust detection the blade is unclutched and rotates freely about its axis until the braking process is activated. The blade rotation is then slowed down and stopped with a braking system. Once the gust is gone, the initial pitch angle is restored with the regular pitch control mechanism.
The adaptation process described above was compared with a pitching mechanism working with the speed of 6 deg/s. It is observed that the unclutching process (semi-active), with the average rate of ca. 26 deg/s, is faster than the pitching mechanism (active solution), cf Fig. 10.

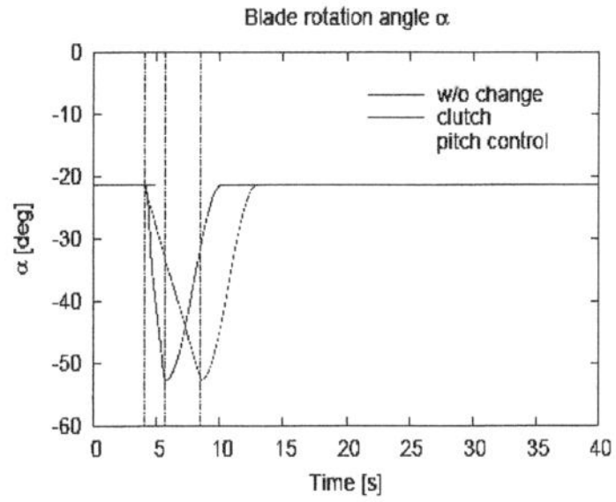

Figure 10: Semi-active vs active device response

The more sudden the gust, the faster the semi-active solution as compared to the pitching mechanism. Fast reaction time creates a possibility to effectively reduce the internal forces resulting from an extreme gust load. This, in turn, could be crucial in the up-scalling process as the blade root bending is an important design criterion. An example answer, i.e. tower and blade bending are depicted on Fig. 11 and Fig. 12 respectively. Results are shown relative to the steady state response.

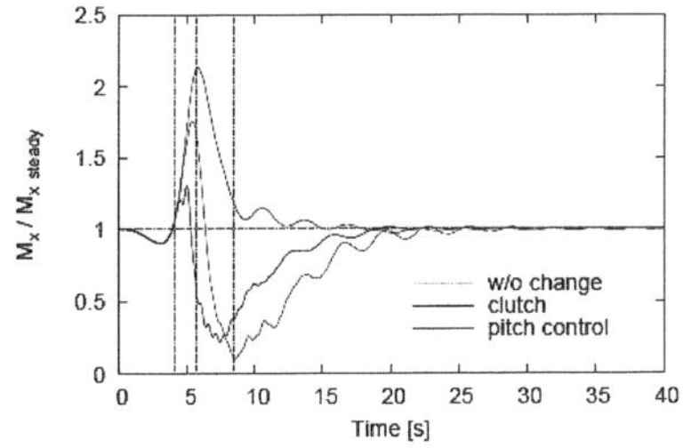

Figure 11: Tower bending moment

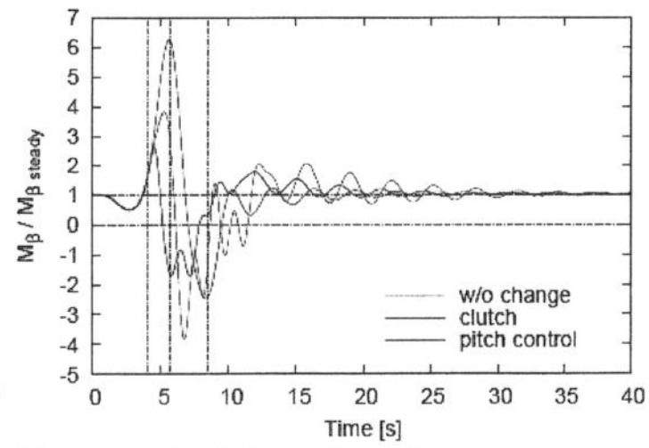

Figure 12: Blade bending moment

Parameters that influence the response are the time instant of unclutching, duration of the free rotation phase, the braking force control and the time delay before initial pitch restoring. Parametric studies have been made for the influence of above variables on the out-of-plane bending moment and results are shown in Figs. 13 and 14.

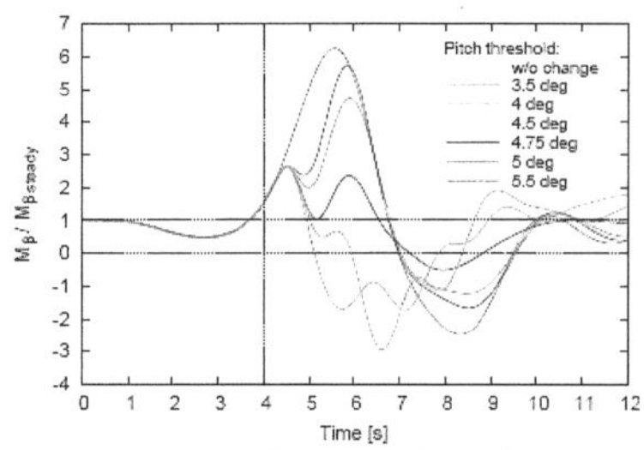

Figure 13: Various free rotation phase duration

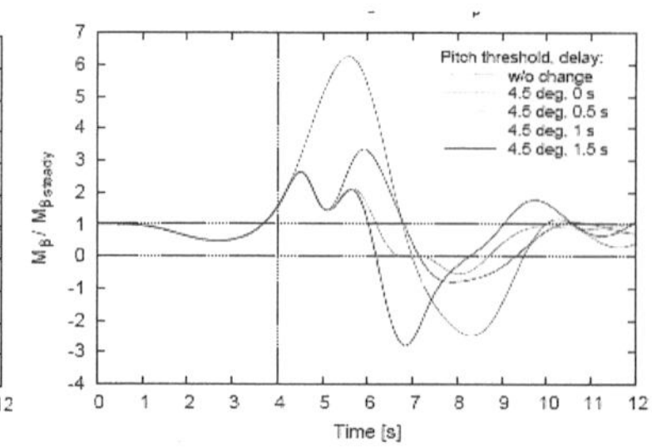

Figure 14: Various initial pitch angle restoring instants

While in the reference case the bending moment range of change is eight-fold its steady state value, the proposed solution can reduce the range of change to three-fold the steady state value.

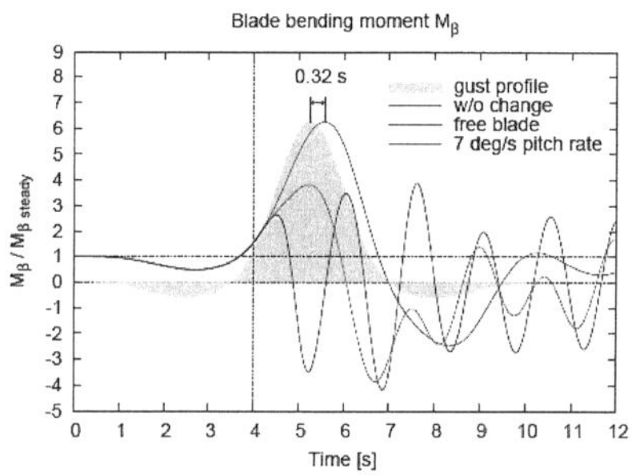

Figure 15: Phase shift between the gust and bending moment maxima

In the reference case the phase shift between the maximum wind speed and the bending moment peak is about 0.3s, which gives an adventage of the direct sensing (wind speed, wind pressure) over the response sensors (stress, strain) (cf. Fig. 15).

Conclusions from the carried out numerical simulations show that the proposed semi-active solution could effectively mitigate the internal forces caused by extreme wind gusts, in particular the blade root bending moment and provides response faster than the pitching mechanism.

Acknowledgement

The authors would like to gratefully acknowledge the financial support through the Polish research project: MAT-INT, PBZ-KBN-115/T08/2004 (2005-2008)

References

[1] Holnicki-Szulc, J. (Ed.), *Smart Technologies for Safety Engineering*, J.Wiley, 2008

[2] J. Holnicki-Szulc, A. Mackiewicz, P. Kolakowski. *Design of Adaptive Structures for Improved Load Capacity.* AIAA Journal vol.36, No.3, 1998.

[3] Holnicki-Szulc J., Pawlowski P., Wiklo M., *High-performance impact absorbing materials - the concept, design tools and applications*, Smart Materials and Structures, vol.12, number 3, 2003

[4] Graczykowski C., Holnicki J., *Protecting offshore wind turbines against ship impacts by means of adaptive inflatable structures*, Shock and Vibration, IOS Press, in print

Solid State Phenomena Vol. 154 (2009) pp 195-201
online at http://www.scientific.net

Active control of landing gear shock absorber characteristic using magnetoreological fluids.

ZBIGNIEW SKORUPKA,
R. KAJKA, R. HARLA, W. KOWALSKI, M. PARAFINIAK,
M. KACPRZAK, G. BALCERKIEWICZ, J. PACHOLSKI, W.LECHNIAK
Instytut Lotnictwa, Al. Krakowska 110/114, 02-256 Warszawa
zk.skorupka@ilot.edu.pl

keywords: landing gear, shock absorber, MRF, smart materials

Smart materials are being used in much larger scale in mechanical solutions. Aviation usage of these materials seems to be natural because of interest in new technologies use in this industry. In this article authors discuss characteristics of magnetoreological fluids as a smart materials, examples of its industrial usage, requirements on landing gear characteristics, design and lab tests of model shock absorber in which MRF was used as damping fluid.

1. Magnetoreological fluid.

Magnetoreological fluids (MRF) are qualified as smart materials, which are controlled by magnetic field. Every MRF consists of two fractions liquid and solid. Liquid fraction is mostly an oil (can be both synthetic or natural), which is responsible for MRF density when magnetic field is not applied. Viscosity of liquid fraction also has also limited interaction with speed of MRF response to magnetic field. Solid fraction of MRF is basically particles of ferromagnetic material mixed with liquid fraction. Solid fraction can be made in different sizes and different susceptibility to magnetic field. Generally MRF because of the composition (mixture of oil and ferromagnetic particles) has the tendency to sedimentation. Depending on the substance used as liquid fraction and the grains sizes of solid fraction sedimentation can be different in the same period of time and be less or more permanent. Some MRF can sediment in such manner that later mixing is impossible what makes their use in constructions (e.g. airplane shock absorbers) impossible.

The change of the property of the MRF consists in the change of the solid fraction particles arrangement in surrounding them liquid fraction. When no magnetic field is applied, the particles of solid fraction are arranged randomly in liquid fraction. When magnetic field is applied particles try to arrange according to lines of the magnetic field creating long chains (fig. 1, fig . 2). This results in the increase of density and shear strength.

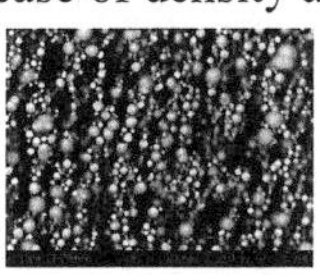

Fig.1 Picture of the particles of the ferromagnetic substance, after applying the magnetic field (picture: http://www.matint.pl/materialy-magnetoreologiczne.php)

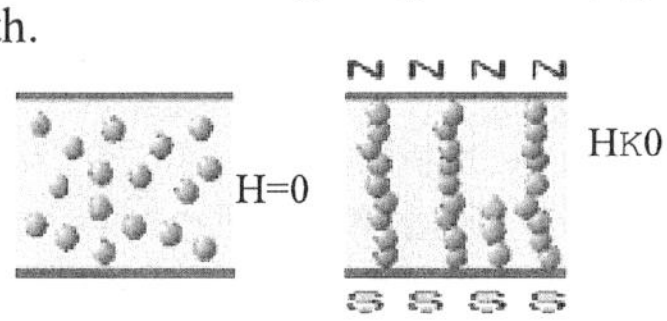

Fig.2 Pattern of the magnetic field influence on the MRF. Dots represents ferromagnetic particles, background represents fluid. (source of picture as in Fig.1)

Magnetic field, which is used to control the MRF, can be generated by any source from the permanent magnet to the coil. Controlling the behavior of MRF follows through the magnetic field value change. MRF reaction to changes in the magnetic field is immediate and allows to use MRF in applications where short times of the reaction are required.

MRFs are offered as the trade product by many companies. The largest and the most well known manufacturer of the MRF and MRF based devices is LORD Corporation.

2. Use of magnetoreological fluid.

Unique properties of MRF create the possibilities of application in actively controlled energy dissipation devices. Example applications of MRF include:
- car suspension systems (fig. 3) ex. vibration dampers
- in pneumatic systems as the speed and position controller
- vibration damping and stiffness systems car seats.
- devices reducing the results of strong wind and quakes (fig. 4)
- in artificial limbs for improved comfort.

Fig. 3. Example MRF based automotive shock absorber (picture: Lord Corp.)

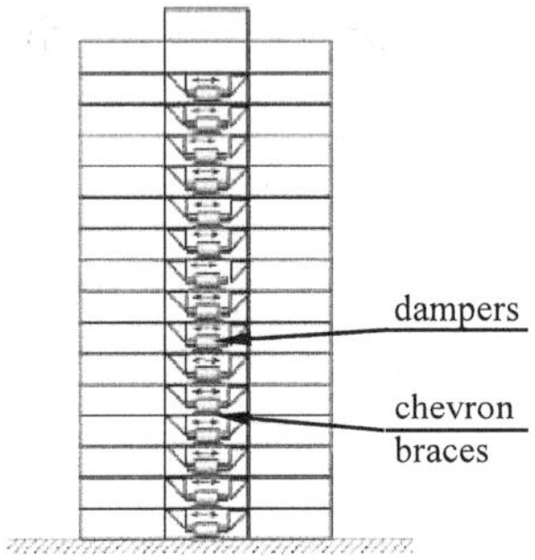

Fig. 4 Example of building seismic vibration damping (picture: Lord Corp.)

3. Requirements for airplane landing gear and shock absorbers.

The main functions the landing gear are:
- to absorb the kinetic energy of the vertical velocity
- to provide elastic suspension during taxiing and ground maneuvers
- to assure safety and comfort for passengers and transported goods during ground maneuvers, start and landing of the airplane

These requirements are fulfilled in landing gear by properly designed shock absorbers and correctly chosen wheels, with the pneumatic part of the wheel helping in the energy dissipation process. Certainly other parts of landing gear are important in order to maintain its integrity, but their energy absorbing properties are not so important.

There are some different landing gear designs that fulfill above criteria. In light airplanes dissipation of the energy can be achieved in different ways:
- spring L/G (in aeronautics L/G is an abbreviation for landing gear) made as a spring beam is both shock absorber and L/G strut (fig. 5)
- flexible elements the most often rubber or different elastomers built-up in the L/G structure.
- steel ring springs
- oleo-pneumatic shock absorbers (fig. 6)

Fig. 5 Spring beam L/G for Cessna 195 airplane (photo: Wikipedia)

Fig. 6 M-28 „Skytruck" main L/G with oleo-pneumatic shock absorber (design Institute of Aviation L/G Department, photo: IA archive)

The first three solutions are preferred in light airplanes because of their low cost in connection with high efficiency rates and low weight. Yet use of the oleo-pneumatic shock absorber is the most effective solution.
Because of the largest efficiency in energy dissipation use of oleo-pneumatic shock absorbers is general in military and commercial airplanes where cost of the construction is not the most important criterion.
Oleo-pneumatic shock absorber absorbs energy by “pushing” a volume of hydraulic fluid against volume of gas (usually nitrogen but can be dry air.)

Oleo-pneumatic shock absorbers carry out two functions:
- a spring or stiffness function, which provides the elastic suspension by the compression of a gas volume.
- a damping function, which dissipates energy by forcing hydraulic fluid through one or more small orifices.

4. Expectation of MRF as shock absorber fluid
MRF because of its consistency is suitable as the replacement of hydraulic oil normally used in shock absorbers. That change carries necessity of certain; quite serious changes in current shock absorbers or design new device initially capable of MRF use. In the case of the modernization of the older solution, there should be throw special attention to different consistency of magnetoreological fluid. Also existence of solid fraction (iron particles) can be destructive for shock absorber parts (particularly seals). Another necessary change in shock absorber is to add one or more coils for MRF control. Design of the new shock absorber eliminates these problems, however this is not always possible.
Main goal of MRF use in airplane landing gear is possibility to correct shock absorber characteristics in order to achieve optimal energy dissipation in every landing condition. Naturally obtaining of the ideal profile is not possible for every landing because of general shock absorber design limits. Range of the possible LG control is also limited by the safety reasons especially by requirement that landing must be safe even during power supply failure.

5. MRF based shock absorber design
MRF based landing gear shock absorber was created by Warsaw Institute of Aviation Landing Gear Department within the project PBZ-KBN-115/T08/04 titled “Metallic, ceramic and polymer smart materials (design – obtainment – properties – use)”. MRF based shock absorber is a developed version of existing I-23 “Manager” small airplane front L/G. Necessary modifications were made to the existing design in order to use MRF but modifications didn’t change L/G layout.

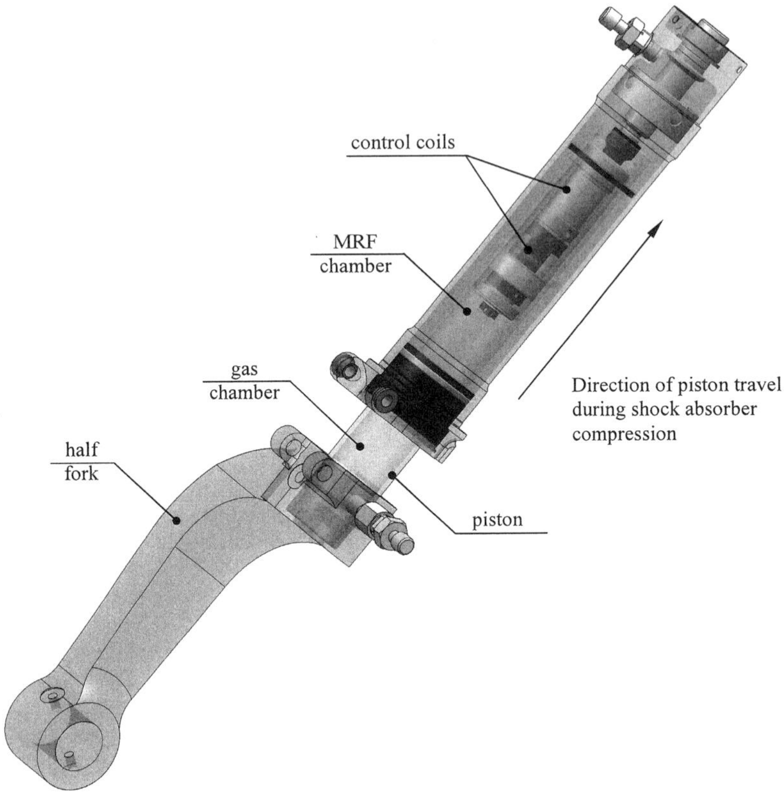

Fig. 7 3D shock absorber model (picture: IA archive)

First stage of the design change was to build MRF control coils. Coils were calculated using Finite Element Method (FEM) by the FEMM [4] software.
After preliminary calculations of coils, which were crucial for shape of the rest of the design, it was 3D modeled. The design process was made by the use of the SolidEdge [5] software for 3D (fig.7) modeling and flat drawing. 3D modeling software was used in order to avoid structural mistakes and to speed up design process.
Shock absorber design process was aided by strength analysis using Finite Element Method (FEM) software. That kind of software significantly speed up design process and gives the chance of optimization before actual manufacture of the product. FEM is widely used in aviation industry as a certified method of strength analysis.

Finally prototype of MRF shock absorber was built. LORD MRF-336AG was used in design as a working fluid. Designers decided to use this MRF because of solid fraction particles size and silicone oil as liquid fraction. Chosen MRF fulfills hydraulic oil flow requirements, easy mixing after sedimentation.

6. MRF based shock absorber laboratory tests

The tests of MRF shock absorber took place in Institute of Aviation Landing Gear Department Laboratory. Both static and dynamic test were made. Static tests were conducted in order to make shock absorber static characteristic. Dynamic tests were made to check shock absorber behavior in life like conditions.

Static tests were conducted at 40T hydraulic press (fig. 8), dynamic tests were made on 10T drop machine (fig. 9).

Fig. 8 40T press for L/G static tests. (picture: IA archive)

Fig. 9 10T drop machine for L/G dynamic tests (picture IA:archive)

During static tests shock absorber was optimized in order to obtain the best possible profile for MRF flow control.

The dynamic tests of the shock absorber were executed on the object which static profile is showed on the graph below (profile Pr_810, fig. 10)

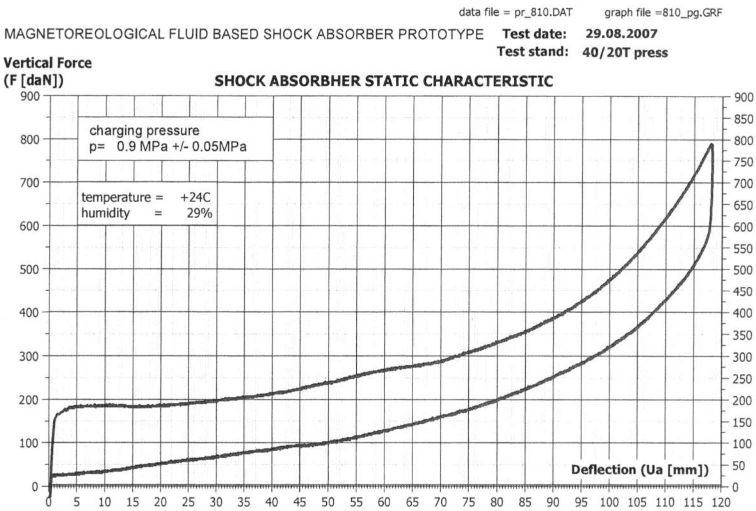

Fig. 10. Static characteristic of MRF shock absorber (picture: IA archive))

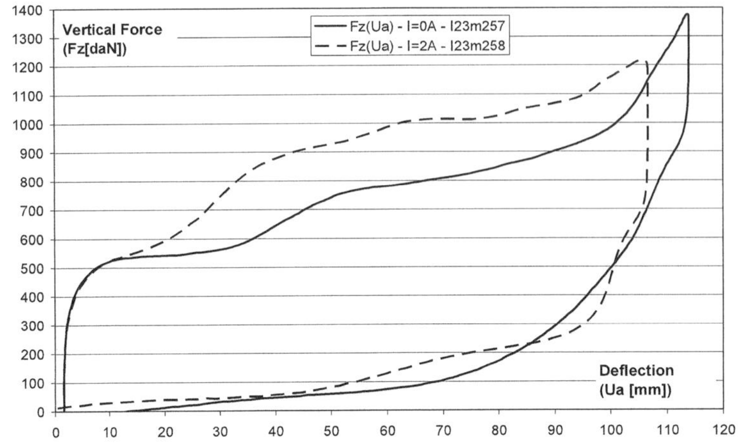

Fig 11. Vertical force versus shock absorber deflection for dynamic tests made with two control currents (both tests made with the same energy). (picture: IA archive)

Dynamic test of L/G is a free drop of defined mass from fixed height in order to achieve desired landing energy.
Dynamic tests I23m257 (fig. 11) and I23m258 (fig. 11) were executed for the same reduced mass and L/G fall (the same energy), with relief of 2/3 dropped mass weight. During first dynamic test coils were not powered, current I=0[A]. In this case (I=0[A]) shock absorber design provide only hydraulic damping by pumping MRF through damping holes as in classic shock absorber solution. Shock absorber hydraulic damping was optimized in order to achieve the same damping value as in standard (non MRF based) I23 shock absorber.
During second dynamic test coils were powered by the current I =2[A]. Maximal L/G load decreased from 1376 [daN] in case of lack of current to 1218 [daN] in case of the 2[A] powered coils. There was about 11,5% decrease in load compared to the hydraulic damping itself (I=0[A]). Load change is a result of increased damping without change in other parameters of dynamic test.

7. Summary and conclusions

Design of MRF based is no more complex than today's standard solutions. There is no need of using additional control system because every modern airplane is equipped in on board power sources and computers that can be used as control electronics.
11,8% load reduction was achieved as the result of MRF use in shock absorber. Reduction was enough to conclude that shock absorber designs based on MRF could be made in future. Achieved effect was promising in spite of current was constant during drop tests. Active control of current can enlarge desired effect.
Use of proposed design can result in making, more reliable and better fatigue resistant landing gears. However high cost of MRF used during tests can put design to the economical test. But when one look at long term advantages of MRF based L/G such as improved safety and comfort designer should consider MRF based design as future of landing gear system.

Literature

[1] Z. Wołejsza, W. Kowalski, A. Lafitte, G. Mikułowski, L. Remmers: State of the art in landing gear shock absorbers. Transactions of the Institute of Aviation, N° 181, Warszawa 2005

[2] Ławniczak A, Milecki A.: Ciecze elektro i magnetoreologiczne oraz ich zastosowanie w technice. Wydawnictwo Politechniki Poznanskiej, 1999.
[3] Lord Corporation web materials: www.lord.com
[4] Finite Elements Method Magnetics web materials: http://femm.foster-miller.net/wiki/HomePage
[5] Siemens PLM Software Solid Edge web materials: http://www.plm.automation.siemens.com/en_us/products/velocity/solidedge/index.shtml
[6] G.Mikułowski, J.Holnicki,: Adaptive landing gear concept-feedbeck control validation, Smart Materials and Structures, 2007, 16, pp. 2146-2158.
[7] N. Currey: Aircraft Landing Gear Design: Principles and Practices, AIAA Education Series, 1989.

Author Index

Keyword Index